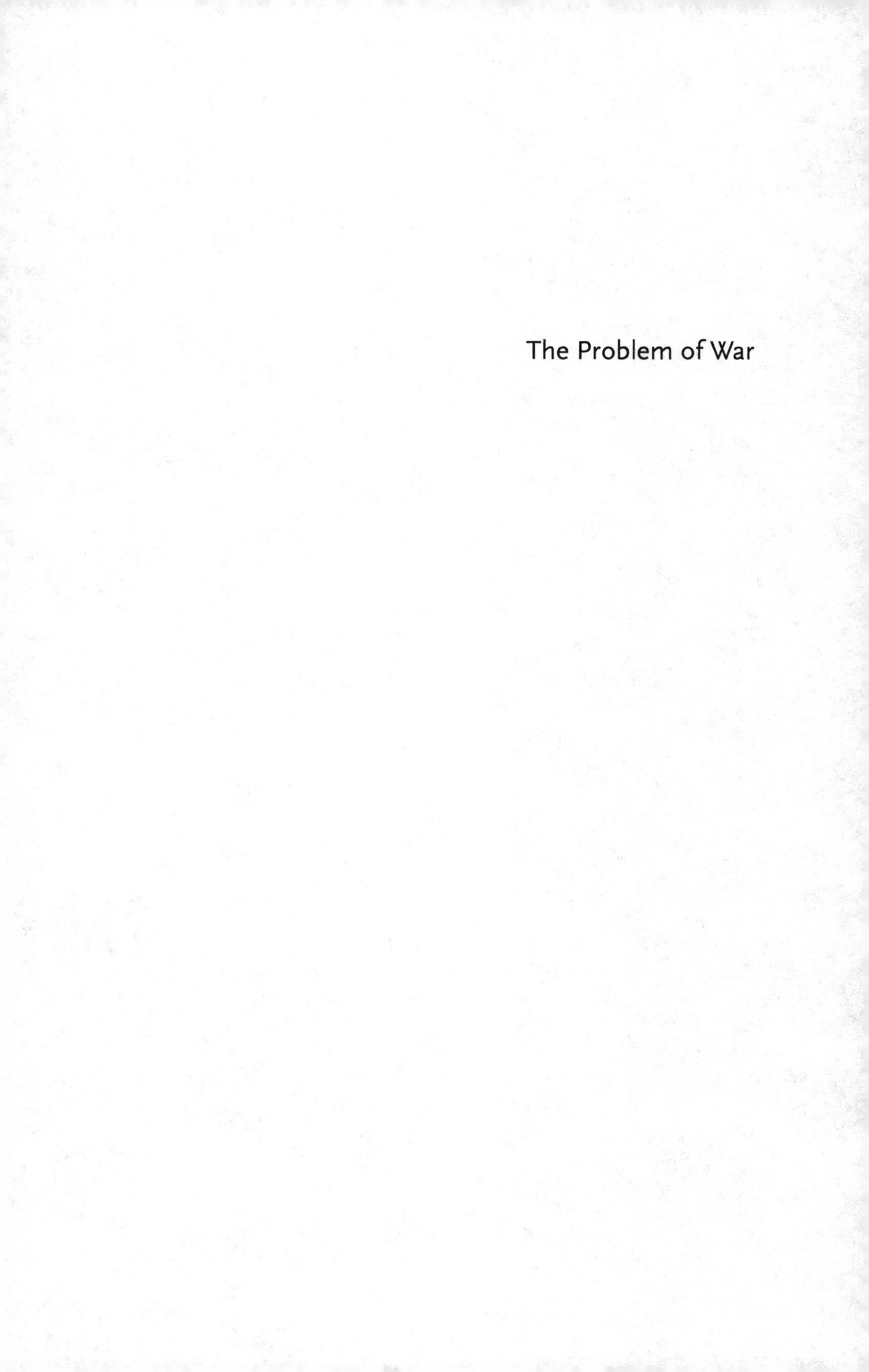

The Problem of War

The Problem of War

DARWINISM, CHRISTIANITY, AND THEIR BATTLE TO UNDERSTAND HUMAN CONFLICT

MICHAEL RUSE

OXFORD
UNIVERSITY PRESS

Oxford University Press is a department of the University of Oxford. It furthers the University's objective of excellence in research, scholarship, and education by publishing worldwide. Oxford is a registered trade mark of Oxford University Press in the UK and certain other countries.

Published in the United States of America by Oxford University Press
198 Madison Avenue, New York, NY 10016, United States of America.

Library of Congress Cataloging-in-Publication Data
Names: Ruse, Michael, author.
Title: The problem of war: Darwinism, christianity, and
their battle to understand human conflict / Michael Ruse.
Description: New York, NY : Oxford University Press, [2019] |
Includes bibliographical references and index.
Identifiers: LCCN 2018012153 | ISBN 9780190867577 (hardcover) |
ISBN 9780190867591 (epub)
Subjects: LCSH: Social Darwinism. | War—Religious aspects—Christianity. |
War—Psychological aspects. | Darwin, Charles, 1809–1882—Influence.
Classification: LCC HM631.R87 2019 | DDC 303.4—dc23
LC record available at https://lccn.loc.gov/2018012153

9 8 7 6 5 4 3 2 1

Printed by Sheridan Books, Inc., United States of America

To the memories of
Regimental Quartermaster Sergeant William J. Ruse, my great-grandfather;
of Staff Sergeant William A. Ruse, my grandfather;
and of William R. E. Ruse, my father.

In Flanders fields the poppies blow
Between the crosses, row on row,
That mark our place; and in the sky
The larks, still bravely singing, fly
Scarce heard amid the guns below.

We are the Dead. Short days ago
We lived, felt dawn, saw sunset glow,
Loved and were loved, and now we lie
In Flanders fields.

Take up our quarrel with the foe:
To you from failing hands we throw
The torch; be yours to hold it high.
If ye break faith with us who die
We shall not sleep, though poppies grow
In Flanders fields.

CONTENTS

PROLEGOMENON

TOWARD THE END OF 1859, the English naturalist Charles Robert Darwin published *On the Origin of Species by Means of Natural Selection, or the Preservation of the Favoured Races in the Struggle for Life.* In that book, Darwin did two things. He argued for the fact of evolution, that all organisms are descended from just a few forms by natural processes, tracing what he called a "tree of life." He argued for a particular force or mechanism of change, natural selection brought on by the struggle for existence. Today's professional evolutionists are committed Darwinians. They think that natural selection is the chief cause of change, and they use this conviction to great effect, explaining organisms right across the spectrum, from the smallest microforms like *E. coli* to the largest of them all: whales, dinosaurs, giant redwood trees. If ever there was an example of what Thomas Kuhn called a "scientific revolution," with Darwinian selection at the center of a successful "paradigm," this is it.

Yet, all is not happy in the Darwinian world. The very idea of evolution is under attack from the religious right, generally evangelical Christians. Even where some kind of change is allowed, it is felt that Darwinian selection can play at best only a minor role. From the so-called Scientific Creationists of the 1960s and 1970s, who denied that the Earth is more than six thousand years old and who claimed that all organisms were created miraculously in six days, to the Intelligent Design Theorists of the 1990s who insist that history is guided by an intelligence from without, Darwinian evolutionary theory is denied and held up for scorn (Numbers 2006; Dembski and Ruse 2004). It is felt that Darwinism is not just wrong, but in some sense deeply disturbing, off on the wrong moral foot as it were. What is truly amazing and upsetting is that this kind of thinking has permeated the world of academia, especially that of the philosophers. Eminent practitioners deny Darwin, and some even go as far as to doubt evolution itself (Plantinga 1991, 2011; Nagel 2012; Fodor and Piattelli-Palmarini 2010). People who pride themselves on their brilliance and commitment to rationality turn away from one of the most glorious jewels in the crown of science. Expectedly, in all of this, it would not be too much to say that there is not only rejection but an odor of hatred, spiced, as is so often the case in these matters, with a good dash of fear.

This is the puzzle that lies behind this book. Why do people react to what is after all a robust empirical science? Why this rejection and nigh loathing? People, including religious people, do not feel this about other sciences. Except perhaps undergraduates who have to pass the test in order to get into medical school, no one harbors the same hostility toward organic chemistry. What is it about Darwinism that is different? It cannot be simply a question of going against the Bible. The sun stopped for Joshua, yet no one denies the heliocentric theory of the universe. My solution is simple. Darwinism is not just an empirical science. It is also a religion, or if you like, a secular religious perspective. It speaks to the ultimate questions, matters of value and importance, as do other religions. As is the case with religion—think Thirty Years' War, think Muslims and Hindus on the Indian subcontinent after partition, think Northern Ireland—those who do not accept Darwinism in this sense react as religious people are wont to do toward rivals. Denial, loathing, fear. This all happens because Darwinism is not just an empirical science. "Not just!" Let me make it very clear that I am not in any way denying that there is a side to Darwinian thinking that is a professional science (Ruse 1988, 2006). For convenience, I shall refer to this as "Darwinian evolutionary theory." It exists and it is very good science. I am claiming that there is (and always has been) another side to Darwinian thinking, and this is religious or akin to religious. For convenience, I shall refer to this, without qualification, as "Darwinism"—rather than "Darwinizing" or some other, rather ugly, made-up word.

This book is part of an ongoing project. In one earlier book, *The Evolution-Creation Struggle* (2005), I argued that Darwinism and Christianity are both in their ways religions, sharing millennial hopes and aspirations, differing in that Darwinians are postmillennialists—thinking we ourselves must make paradise here on earth—whereas Christians (notably those who oppose evolution) tend to premillennialism—thinking that we must prepare for the Second Coming, when God will put all right. Progress versus Providence. In another, more recent book, *Darwinism as Religion: What Literature Tells Us About Evolution* (2017a), I argued, using poetry and fiction as my empirical base, that from the appearance of the *Origin*, Darwinism has been used to speak to such religious issues as God, origins, humans, race and class, morality, sex, sin, and salvation. Speaking to such issues in direct conflation with what Christians have had to say on these topics. Now, I want to take up the problem one more time, through a case study. I want to show how Darwinism and Christianity deal with the topic at hand; how for all of their differences they were (and are) engaged in the same enterprise; that this enterprise is essentially religious: why therefore the hostility felt toward Darwinism—and let us be fair the hostility felt by Darwinians toward Christianity—are nigh inevitable consequences.

My case study is war. This is a promising choice because—apart from the fact that I have not discussed it at all in my previous books, so no built-in biases—both Darwinians and Christians have written a great deal about war. There is a lot of material to try out my claim about the nature of Darwinism. More than this, the two sides tend to take very different positions on the nature and causes of war, these go straight to the root philosophies or theologies of the two systems, and partisans on one side have often made critical comments about those on the other side. This dispute engages us and seems still pertinent today. Near the beginning of this decade the eminent British academic, Nigel Biggar, Regius Professor of Moral and Pastoral Theology at the University of Oxford, published *In Defence of War* (2013). Working from a conventional Christian perspective, he offered a full analysis of the nature of war and its likelihood. He also delved deeply into the moral issues surrounding war. In the course of his discussion, he took a swipe at a book published just a year or two earlier, by the no-less-eminent American academic, Steven Pinker, Johnstone Family Professor in the Department of Psychology at Harvard University. Biggar chose his target well for, working from a conventional Darwinian perspective, in *The Better Angels of Our Nature* (2011) Pinker offered an alternative picture, a full analysis in a Darwinian mode of the nature of war and its likelihood. He also delved deeply into the moral issues surrounding war. Exactly the inquiry of Biggar, and yet chalk and cheese. The two pictures could not be more different. War is a promising choice to try out my hypothesis about the nature of Darwinism.

I appreciate that the reader might be feeling a sense of unease at this point. One would hardly say that war is too sacred a topic to be merely a case study for an inquiry about something else. With reason, however, one might feel that war is simply too big an issue—moral, political, psychological, theological, and more—to be treated in such a way. Any discussion of war has to treat of it entirely in its own right. Let me say two things in response to this very natural and decent feeling. First, above all, mine is a morally driven inquiry. You may think this sounds strange and insincere, for I admit candidly that there are important issues about war that I shall ignore or at least treat in a fashion that many might find inherently inadequate. For instance, there is today a heated discussion about whether war was always part of our heritage. I note with much interest that there is this debate because it feeds right into what I want to say about war and religion. However, while it would be hard to think of a topic of more importance, in the main part of my discussion I am not truly concerned about who is right. This is not only because I am a historian and philosopher and feel no great competence to judge, but because in important respects it is not my topic. I am worried about the nature of Darwinian theory and why so many reject it. I think this a matter of great moral concern—one to which I have devoted much time and effort, writing books, giving talks, and even appearing as a witness for the ACLU in a federal court case. I care about science and its integrity, realizing that it is not just Darwinian thinking under attack, but relatedly claims about global warming, about the need for vaccination, and ways to improve the world's food supplies. I will not rest while one child in the Third World goes to bed hungry because well-fed people of the West have, on entirely bogus philosophical and theological grounds, taken up arms against modern methods of agriculture. I need no lecturing on whether I am approaching war in an appropriate manner.

This said, my second point is that I have not chosen this topic of war lightly or casually. As I write, we are noting the hundredth anniversary of one of the worst of all conflicts, what was then known as the Great War and more recently as the First World War. I was born in England in 1940, at the beginning of the Second World War; but, although important, it was always the First War that haunted my generation—the crippled, crazy, so-sad men, wandering the municipal parks; the lonely, unmarried women, often our primary-school teachers; the photos in the parlor of long-dead, young soldiers, looking so proud in their new uniforms; the powerful literature, taking one down into the trenches with the bombardment above and the mud below; the ubiquitous monuments starting with the Cenotaph in Whitehall. Remembrance or "Poppy" Day, marking the Armistice in 1918 at the eleventh hour, on the eleventh day, of the eleventh month.

Compounding my emotions is the fact that I come from a line of professional soldiers, in the same regiment stretching back to Queen Anne. Both my great-grandfather and my grandfather fought in the First World War, the latter being gassed on the Somme. Breaking with family tradition, although he went to Spain to fight in the Civil War, in the Second World War my father was a conscientious objector. He did contribute by working with prisoners of war. I was raised a Quaker. War, its nature and causes, was a matter of ongoing discussion in my family and of huge concern to me as a child and an adolescent. Then, when twenty-two, I moved to Canada, and realized how important war—particularly the Great War—was to the consciousness and definition of that country. Vimy Ridge is where the Canadians, at Easter 1917, for the first time fighting together as a group, took a hill that had hitherto been impregnable. It has the iconic status of Gallipoli for Australians and New Zealanders. If that were not enough, every workday for thirty-five years, I walked coming and going past the birthplace of Colonel John McCrea, author of "In Flanders Fields," the most famous poem from that war. I could talk more about my time as a graduate student in the USA during the build-up of the Vietnam War, but the point is made. I want to understand Darwinism and its status, but I am doing so through a case study that has for me great emotional and moral significance.

Finally, let me say, in a non-sanctimonious way, that I have never seen the point of being a scholar just for the sake of being a scholar. I want to make a difference to the world. Hence, at the end of my main discussion, I shall pull back and see if what I have been able to find and explain does have serious implications for how we move forward next, both beyond the hostility between science and religion broadly construed and in our particular case toward an understanding of war. I have been able to do little more than sketch out a few thoughts. So perhaps it is better to think of my final chapter less as the conclusion to this book and more as the opening of another book. That said, I believe there are findings of huge importance both for Darwin scholars and for war theorists—findings I had not anticipated when I set out on this project. It is for this reason, wanting things to emerge in their own right rather than because they were injected surreptitiously, I have striven to let people speak in their own words rather than mine.

In writing this book, I am hugely indebted to four men. The first is Peter Ohlin, my editor at Oxford University Press. He has encouraged me and supported me in this project, and through his own wise comments and even more through the referees he carefully chose has helped me to shape the book and to address the issues that ought to be addressed, and (as important) to ignore those matters that are not strictly relevant to my main theme. Second, Robert Holmes, my teacher of moral philosophy over fifty years ago at the University of Rochester. Looking back, and as shown by our subsequent interests, I don't think either of us really enjoyed the analytic approach to philosophy that was (and in many places still is) *de rigueur*. He let me write a term paper on punishment, the kind of exercise in those days looked down upon as near casuistry. Through this I sensed that, along with looking at science, value inquiry is all-important for the philosopher. Third, my colleague here at Florida State University, John Kelsay in the Department of Religion. He is both an expert on just war theory in Islam and, as a Presbyterian minister, understands the thinking of John Calvin to the extent that his nonbelieving friend worries that it might not be entirely healthy. He has been a constant mine of information and ever willing to cast a critical eye. More importantly, he is a friend who believes in my project. We team-taught a graduate course on war. It was for me a great privilege to work with a man of such intelligence and integrity.

Fourth, thanks go to a man whom I do not think I have ever met, the Australian historian Paul Crook. I regard his *Darwinism, War and History* as one of the most important books ever written on Darwinian thought and its influence. So impressed am I by that book that I hesitated to move in and write anything else on the topic. I realize, however, that Crook is writing as a historian, trying to show (very successfully) the full range of thinking about war in the fifty or so years after the *Origin*. I am a philosopher and my topic—as I have explained—is the understanding of Darwinism today. I am not writing in what historians call a "Whiggish" fashion, trying to show that everything led progressively up to the present. I am an evolutionist, and I think that the present is best understood by looking at the past. This is why I feel able to write my book alongside that of Crook and why, as need be, I feel able to make full and explicit use of his findings and interpretations.

I want also to pay homage to the memories of Lucyle and William Werkmeister, whose generous legacy to the Philosophy Department at Florida State University supports my professorship and made it possible to take students on a trip to Flanders, to the battlefields of the First World War. The Somme, especially Delville Wood, Ypres, and Passchendaele, which are of such poignant, historical importance. Truly, one of the most important and humbling experiences of my life. My research assistant and friend, Jeff O'Connell, has been his usual quiet and efficient self. As always, I declare my love for Lizzie. Together with my students, my wife and family—not forgetting the dogs!—have ever been a bright beacon in my life, making it all worthwhile.

The Problem of War

1 | Darwinian Evolutionary Theory

CHARLES ROBERT DARWIN (1809–1882) WAS the son of a physician; grandson on his father's side of the physician and evolutionist, Erasmus Darwin; and on his mother's side of Josiah Wedgwood, founder of the pottery works that still bears his name (Browne 1995, 2002). On his way to becoming a full-time naturalist, Darwin spent five years on HMS *Beagle*, circumnavigating the globe. The experiences on this trip spurred his becoming an evolutionist. He published his great work, *On the Origin of Species* (1859), when he had just turned fifty, and followed it up twelve years later by a work on our species, *The Descent of Man and Selection in Relation to Sex* (1871). As a young man, Darwin had been raised as a fervently believing member of the Church of England. Around the time of becoming an evolutionist, he softened into a kind of deism, God as Unmoved Mover. Late in life he became a skeptic, what Thomas Henry Huxley fashionably labeled an "agnostic." No matter. The English know a hero when they see one. When Darwin died, recognized as one of the very greatest of scientists, he was buried in Westminster Abbey alongside Isaac Newton. Both men have recently appeared on the backs of English banknotes.

Evolution: The Cause

Charles Darwin was a skilled methodologist and the structure of the *Origin of Species* shows this (Ruse 1975, 1979). He himself referred to the book as "one long argument," which one might amend as "one very carefully constructed long argument." Although it is the fact of evolution that looms over his whole enterprise, Darwin begins with the cause of change, natural selection. Always with the Newtonian gravitational theory in mind—had not Immanuel Kant set the challenge for the ambitious young naturalist by declaring that there would never be a Newton of the blade of grass?—Darwin wanted to show that his cause had the status of the Newtonian cause of gravitational attraction. To do this he had to show that selection was what the theorists called a *vera causa* or true cause. The physicist-philosopher John F. W. Herschel (1830) was the authority, saying that a true cause is a force of which one has direct experience. We know that a force is pulling the moon down toward the earth because when we twirl a stone on a string around our finger, we feel the pull of the string as the stone struggles to escape. Darwin, who grew up in rural England, turned to the barnyard for guidance. Breeders—and this also includes pigeon fanciers and the like—select for the characteristics they desire, and over the generations they arrive at the forms, animals, and plants, that they want:

> Youatt, who was probably better acquainted with the works of agriculturalists than almost any other individual, and who was himself a very good judge of an animal, speaks of the principle of selection as "that which enables the agriculturist, not only to modify the character of his flock, but to change it altogether. It is the magician's wand, by means of which he may summon into life whatever form and mould he pleases." Lord Somerville, speaking of what breeders have done for sheep, says:—"It would seem as if they had chalked out upon a wall a form perfect in itself, and then had given it existence." (Darwin 1859, 31)

Now with his hands-on force to back and guide him, Darwin turned to the world of nature. He offered a two-part argument. First, he argued to what was known as a "struggle for existence," although in Darwin's case it was more pertinently a "struggle for reproduction."

> A struggle for existence inevitably follows from the high rate at which all organic beings tend to increase. Every being, which during its natural lifetime produces several eggs or seeds, must suffer destruction during some period of its life, and during some season or occasional year, otherwise, on the principle of geometrical increase, its numbers would quickly become so inordinately great that no country could support the product. Hence, as more individuals are produced than can possibly survive, there must in every case be a struggle for existence, either one individual with another of the same species, or with the individuals of distinct species, or with the physical conditions of life. (63)

Selection cannot take place unless there is something on which it can work, differences between organisms—variations. Darwin was convinced that there is a constant supply of new variations always appearing in any population, domestic or wild. He had little idea about precisely why there are such variations—later he was to devise a theory (never introduced into the *Origin*) about small particles or "gemmules" being collected in the sex cells from all over the body (Darwin 1868; Olby 1963)—but on one thing Darwin was always firm. However variations may be caused, they do not occur to order or to satisfy need. There is no direction to variation; it is not guided by natural or supernatural forces. In this sense, it is "random," meaning not that it is uncaused—Darwin was sure that there were causes—but that it does not appear to order.

With this, Darwin was ready for the main force.

> Let it be borne in mind in what an endless number of strange peculiarities our domestic productions, and, in a lesser degree, those under nature, vary; and how strong the hereditary tendency is. Under domestication, it may be truly said that the whole organisation becomes in some degree plastic. Let it be borne in mind how infinitely complex and close-fitting are the mutual relations of all organic beings to each other and to their physical conditions of life. Can it, then, be thought improbable, seeing that variations useful to man have undoubtedly occurred, that other variations useful in some way to each being in the great and complex battle of life, should sometimes occur in the course of thousands of generations? If such do occur, can we doubt (remembering that many more individuals are born than can possibly survive) that individuals having any advantage, however slight, over others, would have the best chance of surviving and of procreating their kind? On the other hand, we may feel sure that any variation in the least degree injurious would be rigidly destroyed. This preservation of favourable variations and the rejection of injurious variations, I call Natural Selection. (80–81)

One absolutely fundamental thing must be grasped right here, for it is the key to the whole Darwinian story. Selection does not merely cause change, but change of a particular kind. It makes for useful features or characteristics, things that will help their possessors in the struggle. It makes for things like the eye or the hand, the leaf or the root, the instinct to build a nest and the fear of the predator. It makes for "adaptations." Change is not higgledy-piggledy, but as if contrived. "How have all those exquisite adaptations of one part of the organisation to another part, and to the conditions of life, and of one distinct organic being to another being, been perfected? We see these beautiful co-adaptations most plainly in the woodpecker and missletoe; and only a little less plainly in the humblest parasite which clings to the hairs of a quadruped or feathers of a bird; in the structure of the beetle which dives through the water; in the plumed seed which is wafted by the gentlest breeze; in short, we see beautiful adaptations everywhere and in every part of the organic world" (60–61). The answer is natural selection.

Evolution: The Fact

With the case made for natural selection, Darwin then started to flesh out his vision of life's history, beginning with his fabled tree of life. A division of labor operates in the world of organisms at least as efficiently as in the world of human industry, and organisms specialize to improve their own chances and to avoid competition. This leads to branching and speciation. "The affinities of all the beings of the same class have sometimes been represented by a great tree. I believe this simile largely speaks the truth" (129). Buds appear and some thrive and flourish, and so branches are formed and the tree fills out and gets ever bigger and higher. "As buds give rise by growth to fresh buds, and these, if vigorous, branch out and overtop on all sides many a feebler branch, so by generation I believe it has been with the great Tree of Life, which fills with its dead and broken branches the crust of the earth, and covers the surface with its ever branching and beautiful ramifications" (130).

I wrote of Darwin as a skilled methodologist. If Herschel was the theoretician of science who took what we might call an empiricist stance toward true causes—we must experience them in some way—the Cambridge historian and philosopher of science William Whewell (1840) was he who took a more rationalist stance to the issue. Drawing particularly on the example of the wave theory of light, which was in the 1830s (the time when Darwin was formulating his theory) conquering over the hitherto-accepted Newtonian particle theory of light, Whewell argued that direct experience is unnecessary. No one sees the waves of light and analogies with physical things like waves in water are very loose. Better to say that we accept unknown forces because of the indirect evidence. As a detective puzzles out the culprit because of the clues—bloodstains, weapon, broken alibi—so we assign *vera causa* status because of the clues, as it were. Gathering together the clues from all over—what Whewell called a "consilience of inductions"—makes for secure status.

Darwin, who knew Whewell well from when he was an undergraduate, took note of this. Significantly, whenever later he was challenged about natural selection, he would bring up the wave theory. In the *Origin*, with selection now a given, Darwin offered a full-blown consilience, going systematically through the life sciences, showing how evolution through selection solves the problems, and conversely is given credibility because of its success. Starting with social behavior, Darwin showed how the hymenoptera succeed because of their in-group interactions, and how such things as the intricate comb of the honeybee are built in such a way as to maximize product with minimal effort. Here, as throughout the *Origin*, Darwin relied heavily on the analogy from the domestic world. Many farm animals are castrated, yet their desired qualities can be passed on indirectly through fertile relatives. Likewise, the sterile workers in nests offer no great problem to the selectionist. In aiding their fertile relatives, their own features get passed to the next generation.

Moving on, the subject matter of paleontology, the fossil record, was readily shown to be a natural consequence of evolution through selection.

> We can understand, from the continued tendency to divergence of character, why the more ancient a form is, the more it generally differs from those now living. Why ancient and extinct forms often tend to fill up gaps between existing forms, sometimes blending two groups previously classed as distinct into one; but more commonly only bringing them a little closer together. The more ancient a form is, the more often, apparently, it displays characters in some degree intermediate between groups now distinct; for the more ancient a form is, the more nearly it will be related to, and consequently resemble, the common progenitor of groups, since become widely divergent. (344–345)

Darwin had first been flagged to the possible need of an evolutionary hypothesis when, in September 1835, he had visited the Galapagos archipelago in the Pacific. The different birds and reptiles on the different islands suggested change that surely had to be natural. No surprise, therefore, that biogeography or geographical distributions got star billing in the *Origin*. How but through evolution through selection could you explain the fact that denizens of the Galapagos are like the nearby South American continent whereas the denizens of the Canary Isles in the Atlantic are like those of the nearby African continent? Turning now to some of the traditional areas of biology, Darwin looked relatively quickly at systematics (taxonomy), morphology, and anatomy. He was writing a century after Linnaeus had imposed order on the classification of organisms, and Darwin was happy to show that evolution through selection gives a full and adequate reason for the kind of branching tree that the Linnaean system presupposes. Likewise, and note how greatly Darwin relies on work done and established by earlier biologists, starting at least with Aristotle, people had long known about the similarities in the skeletons of animals very different, what the anatomist Richard Owen (1848) christened "homologies." It took Darwin to show that these are the legacy of evolution.

Almost at the end with embryology. Here was a real triumph. Well known was the fact that organisms very different as adults—humans and chickens—can have embryos that are virtually indistinguishable. Darwin argued that this points to common descent and that the reason for the similarity of embryos is because selection does not work hard at that level to separate out forms. For support, he turned to the world of the breeders—dogs and horses—expecting that, since breeders work for the desired adult forms, we should expect more similarities in the younger forms. Despite denials by the professionals, almost smugly Darwin responded that his expectations were confirmed both with respect to bulldogs versus greyhounds and with respect to carthorses versus racehorses.

With a quick look at vestigial organs like the appendix, Darwin was now ready to move to make perhaps the most famous declaration in the whole of science.

> It is interesting to contemplate an entangled bank, clothed with many plants of many kinds, with birds singing on the bushes, with various insects flitting about, and with worms crawling through the damp earth, and to reflect that these elaborately constructed forms, so different from each other, and dependent on each other in so complex a manner, have all been produced by laws acting around us.

Continuing:

> Thus, from the war of nature, from famine and death, the most exalted object which we are capable of conceiving, namely, the production of the higher animals, directly follows. There is grandeur in this view of life, with its several powers, having been originally breathed into a few forms or into one; and that, whilst this planet has gone cycling on according to the fixed law of gravity, from so simple a beginning endless forms most beautiful and most wonderful have been, and are being, evolved. (490–491)

"The Descent of Man"

Darwin had formulated his theory by the end of the 1830s, and early in the next decade wrote out versions of it. For reasons that are still not fully understood, he sat on his thinking for nigh twenty years (Browne 1995). In 1858, as his friends were predicting, the very worst happened. A young collector named Alfred Russel Wallace sent to Darwin of all people—obviously, his commitment to evolution was getting out—a short manuscript sketching out virtually the same theory as that which lay unpublished. Hurriedly, Darwin wrote the *Origin* and it appeared the next year. Relieved, Darwin then thought he would spend his years writing individual volumes, detailing the main points of the quick overview offered in the *Origin.* He did complete one such item, a two-volume work on variation under domestication (Darwin 1868). But, basically, Darwin—who was then middle aged, who had well-enough family money to live comfortably, who was a semi-invalid from a still-unknown ailment—wanted a quiet life and to work on little projects that he could do readily from home. He wrote a book on orchids, another on insectivorous plants, and yet a third on earthworms.

Wallace, however, disrupted Darwin's peace, for in the 1860s the younger man took up spiritualism, arguing that human evolution cannot be explained entirely by selection and must involve non-natural forces (Ruse 2012; Richards and Ruse 2016). Darwin was horrified. In the *Origin,* he had wanted to get the main theory on the table before all erupted in a debate about humans—the "monkey" or "gorilla theory" as it was at once called. However, Darwin had always thought that we humans are part of the picture, and indeed from his private notebooks the first evidence that we have that Darwin had grasped firmly the idea of selection is a passage where he not only applied selection to humankind but also argued that intellectual differences are a result of selection. Shades of the 1970s and the row over so-called genetic determinism! However, that he not be accused of cowardice, at the end of the *Origin* Darwin made his position very clear. "In the distant future I see open fields for far more important researches. Psychology will be based on a new foundation, that of the necessary acquirement of each mental power and capacity by gradation. Light will be thrown on the origin of man and his history" (488).

Responding to Wallace, Darwin took up again the pen, and thus appeared the two-volume *The Descent of Man, and Selection in Relation to Sex.* In major respects, there are no great surprises. Humans evolved and natural selection played a crucial role in this process. However, as the subtitle flags us, in this new work Darwin made much of a secondary form of selection, introduced in the *Origin* but then essentially neglected. Sexual selection occurs within species and involves the competition for mates. It has two forms: one through male combat, as when stags evolve horns or antlers to fight others for possession of the females; and one through female choice, as when peahens chose the peacocks with the most dazzling display of tail feathers. Wallace had argued that we must involve forces beyond natural selection because so many human features have no obvious adaptive function—human hairlessness for instance. Unlike most, including Darwin, the lower-middle class Wallace had been forced to live with the indigenous people on his collecting trips, so he knew firsthand that these "savages," as the Victorians called them, were very much our intellectual equals if given the chance. Yet, Wallace could see no good selective reason for this reasoning power if, as he thought, the natives never or rarely used it. Wallace therefore called on spirit forces to fill the causal gap.

Darwin agreed with Wallace that natural selection cannot do the job, but argued that sexual selection can speak to the need. Hence, much of the *Descent* is given over to a general discussion of sexual selection and then, at the end, as it applied to humankind.

> Man is more courageous, pugnacious, and energetic than woman, and has a more inventive genius. His brain is absolutely larger, but whether relatively to the larger size of his body, in comparison with that of woman, has not, I believe been fully ascertained. In woman the face is rounder; the jaws and the base of the skull smaller; the outlines of her body rounder, in parts more prominent; and her pelvis is broader than in man; but this latter character may perhaps be considered rather as a primary than a secondary sexual character. She comes to maturity at an earlier age than man. (Darwin 1871, 2, 316–317)

And so on and so forth. All very Victorian, a point of which I am going to make much in a moment.

Reception—Darwin Demoted

Pull back now and start to put Darwin's work into context. He was not the first evolutionist. His own grandfather Erasmus Darwin had adopted some version of transformation at the end of the eighteenth century and well known was the French biologist Jean Baptiste de Lamarck, who had proposed a version of descent in his *Philosophie Zoologique* published in 1809, the year of Darwin's birth (Bowler 1984). Amusingly, Lamarck has given his name to the (now-known-to-be-false) mechanism of the inheritance of acquired characteristics. In fact, for Lamarck it was always a minor part of the picture and it was in ways more important for Charles Darwin, who was ever a Lamarckian and who became more so over the years. Then, in 1844, just as Darwin was writing down (in private) a full-blown essay on his theory, the Scottish publisher Robert Chambers published the evolutionary tract, *The Vestiges of the Natural History of Creation.* It caused a huge argument and was rejected often—at great length—by the scientific establishment.

What Darwin did was change the status of claims about the very fact of evolution from flaky pseudoscience to respectable hypothesis accepted by almost all. He really was revolutionary in this respect. We all know the famous story of Darwin's great supporter, his "bulldog," Thomas Henry Huxley who, at the 1860 meeting of the British Association for the Advancement of Science, squared off against "Soapy Sam" Wilberforce, high church Bishop of Oxford (and son of William Wilberforce of slave-trade-abolition fame). Supposedly, the bishop asked the professor—Huxley taught at the Royal School of Mines—whether he was descended from monkeys on his grandfather's side or his grandmother's side. To which the response came that he had rather be descended from a monkey than from a bishop of the Church of England. Almost certainly embellished in the telling, but like Moses and the Red Sea, a myth with meaning. One can joke, but, as everyone realized, the ball was over before the guests arrived. Already it was becoming clear that Darwin's case for evolution as fact was all-conquering. That consilience, with all of those neat explanations like the sterile worker bees, was just too powerful and convincing to be ignored or rejected. As just noted, in many respects, Darwin was not in new territory. He was seizing on facts that everyone knew, like the puzzling similarities between our skeletons and those of fish and reptiles, let alone monkeys and apes. He was seizing on them and giving a most convincing explanation. Like the story of the emperor's new clothes, as soon as Darwin said "it evolved," everyone said they had known it all of the time. By the mid-1860s, for instance, students at Cambridge University taking the final examination in biology were told to assume evolution and discuss the putative causes. That Darwin's son Frank got a first-class degree suggests that that debate was over. We know—a reason for writing this book—that many people, particularly in America, never accepted evolution and many still don't, but generally even among church people evolution was a given—at least for non-humans (Roberts 1988). No one cared about the immortal souls of warthogs.

Darwin's mechanism of natural selection, however, was another matter. Loud and influential voices in the professional biological community were not at all enthusiastic about it (Bowler 1983; Ruse 1979a, 1996). Thomas Henry Huxley (1860) was typical, for although he allowed that selection plays some role, overall as a professional biologist he was unmoved by Darwin's mechanism and opted instead for other forces, particularly evolution by jumps—saltationism—fox into dog in one generation. This suggests that he was indifferent to Darwin's main problem, namely explaining adaptation. For Darwin, it was simply not plausible that the eye appear in one generation randomly. At least, not if it is seen as an organ of contrivance with a distinctive purpose. For Huxley, things were otherwise. He was an anatomist—later paleontologist—and tracing similarities between forms, homologies, was his main research focus. Adaptations, like camouflage, tended to obscure deeper relationships. Hence, in Huxley's massive 165 biology lecture series, he would spend but half of one lecture on evolution, and less than ten minutes on natural selection. There were lots on Negro teeth size, however, and other such vital topics.

In France, the evolutionists naturally tended to Lamarckism and in Germany, likewise reflecting national influences, it was German Romanticism—harking back to Johann Wolfgang von Goethe and Friedrich Schelling and Lorenz Oken—that had the big input. Romanticism is essentially Platonic, seeing all as one stemming from the Good, and so there was—notwithstanding the world picture being developmental—much emphasis on connections and similarities and underlying shared patterns. It is little surprise that, in the second half of the nineteenth century, the most famous German evolutionist of them all, Ernst Haeckel, despite calling himself a Darwinian, was basically indifferent to selection and is best known as the advocate of the biogenetic law, ontogeny recapitulates phylogeny. The pattern of individual development finds an echoing response in the history of life (Richards 2008). American biology was more influenced by German thought than by British—increasingly, young Americans crossed the ocean to do the new doctoral degree in a German university. At home, the biggest influence was the Harvard-located Swiss import, the ichthyologist (and discoverer of the Ice Age) Louis Agassiz. A student of both Schelling (a philosopher) and Oken (an anatomist), although he himself never accepted evolution, his many students (including his own son) embraced it readily, but kept the faith in seeing evolution through the lens of Romanticism or (as it was known) *Naturphilosophie*. Natural selection got little billing (Lurie 1960).

At the beginning of the twentieth century, one of the first biology professors at Stanford University was the Kansas-born entomologist Vernon Lyman Kellogg. Kellogg's real strength was in popularization, and his *Darwinism Today* (published in 1905) was a major contribution to that genre. One is flagged at once that things do not bode well for the old naturalist from Down House, in Kent, the long-time home of the Darwins. The first chapter is titled "Introductory: The 'Death-bed of Darwinism.'" It is true that this refers to a pamphlet—in German!—but the tone is set. Natural selection has a role in things, but chiefly as a kind of clean-up process after the important work has been done by new variation. "The living stream of descent finds its never-failing primal source in ever-appearing variations; the eternal flux of Nature, coupled with this inevitable primal variation, compels the stream to keep always in motion, and selection guides it along the ways of least resistance" (Kellogg 1905, 374). Good, but not quite good enough. "Darwinism, then, as the natural selection of the fit, the final arbiter in descent control, stands unscathed, clear and high above the obscuring cloud of battle. At least, so it seems to me. But Darwinism, as the all-sufficient or even most important causo-mechanical factor in species-forming and hence as the sufficient explanation of descent, is discredited and cast down."

So what then is the cause of this all-important new variation? In an American tradition going back to the first evolutionists after Darwin, notably Alpheus Hyatt and Edward Drinker Cope, something much like the Lamarckian inheritance of acquired characteristics seems needed and available (Ruse 1996). Are we not justified in drawing "the logical conclusion that the species change and adaptation is derived, not by the chance appearance of the needed variation, but by the compelled or determined appearance of this variation? In other words when species differences and adaptations are identical with differences and modifications readily directly producible in the individual by varying environment, are we not justified, on the basis of logical deduction, to assume the transmutation of ontogenetic acquirements into phyletic acquirements . . . ?" (382).

In the light of all of this, there is little surprise that today there is a veritable cottage industry of historians of science leaving you to hope that, if indeed Darwin was on the back of a banknote, it was not one of the higher denominations. Entirely typical is the Canadian historian Bernard Lightman (2010) who argues that, as far as the general British and American public was concerned, when it came to evolution, "Darwin was just one author among many competing for their attention and patronage" (20). Little surprise that today's most indefatigable reporter on the history of evolutionary theory, Peter Bowler, has authored a series of books with revealing titles: *The Eclipse of Darwinism: Anti-Darwinism Evolution Theories in the Decades around 1900; The Non-Darwinian Revolution: Reinterpreting a Historical Myth*; and most recently, *Darwin Deleted: Imagining a World without Darwin.* From the amazon.com website:

> *Darwin Deleted* boldly offers a new vision of scientific history. It is one where the sequence of discovery and development would have been very different and would have led to an alternative understanding of the relationship between evolution, heredity, and the environment—and, most significantly, a less contentious relationship between science and religion. Far from mere speculation, this fascinating and compelling book forces us to reexamine the preconceptions that underlie many of the current controversies about the impact of evolutionism.

Reception—Darwin Promoted

Darwin deserves credit for making the fact of evolution convincing. For his mechanism, the story is otherwise. It is true that selection is important today—on this, we shall speak more in a later chapter (8)—but only after it was resurrected by the geneticists in the twentieth century. Although to say that Darwin left it dead is to suggest incorrectly that it was alive in the first place. Natural selection was essentially dead on arrival.

A good story, but another candidate for the Moses-and-the-Red-Sea Prize for the tallest story of all time. A very different picture emerges if you start to broaden your vision of where and how and on whom Darwin and his mechanism of natural selection might have had some effect. Start with the fact that, although among full-time scientists natural selection was certainly not at first a smash-hit success, even from the beginning there were people doing very professional evolutionary biology using natural selection to solve hitherto-unanswerable problems (Ruse 2017a). A sometime traveling companion of Wallace, Henry Walter Bates, working in South America, showed how nonpoisonous butterflies have evolved to mimic poisonous butterflies, thus having adaptive camouflage against predators, birds (Kimler and Ruse 2013). Also in South America, the German-born Fritz Müller likewise came up with a selection-based explanation for other forms of mimicry—poisonous butterflies evolve to look like other poisonous butterflies so that birds will more quickly learn to avoid each and every one of them. The

German biologist August Weismann, as did Wallace, studied other facets of butterfly evolution. Later in the century, the professor of biology at Oxford University, E. B. Poulton, took up animal coloration from a selective perspective, and there was much interest among lepidopterists in so-called industrial melanism, where insects are adapted to the filth deposited on trees by the smoke from industry. "I believe . . . that Lancashire and Yorkshire melanism is the result of the combined action of the 'smoke,' etc., plus humidity [thus making bark darker], and that the intensity of Yorkshire and Lancashire melanism produced by humidity and smoke, is intensified by 'natural selection' and 'hereditary tendency'" (Tutt 1890, 56).

Although he seems not to have reacted—Darwin always thought that selection would take many generations and be unobservable in our lives—one enthusiast even told the old naturalist about this.

> My dear Sir,
>
> The belief that I am about to relate something which may be of interest to you, must be my excuse for troubling you with a letter.
>
> Perhaps among the whole of the British Lepidoptera, no species varies more, according to the locality in which it is found, than does that Geometer, Gnophos obscurata. They are almost black on the New Forest peat; grey on limestone; almost white on the chalk near Lewes; and brown on clay, and on the red soil of Herefordshire.
>
> Do these variations point to the "survival of the fittest"? I think so. It was, therefore, with some surprise that I took specimens as dark as any of those in the New Forest on a chalk slope; and I have pondered for a solution. Can this be it?
>
> It is a curious fact, in connexion with these dark specimens, that for the last quarter of a century the chalk slope, on which they occur, has been swept by volumes of black smoke from some lime-kilns situated at the bottom: the herbage, although growing luxuriantly, is blackened by it.
>
> I am told, too, that the very light specimens are now much less common at Lewes than formerly, and that, for some few years, lime-kilns have been in use there.
>
> These are the facts I desire to bring to your notice.
>
> I am, Dear Sir, Yours very faithfully,
>
> A. B. Farn
>
> Letter from Albert Brydges Farn on November 18, 1878. (Darwin Correspondence Project, 11747)

Then, at the end of the century, in Britain the so-called biometricians, notably the empiricist Raphael Weldon and the mathematician Karl Pearson, started to do sophisticated selection experiments and backed them with no less sophisticated mathematical analyses (Provine 1971; Ruse 1996). Likewise elsewhere. At Cornell University in upstate New York, the entomologist and taxonomist John Henry Comstock, with a vast knowledge of insect pests, a key figure in the development of scientific agriculture, hugely influential thanks to texts that he himself published, writing in 1893, gave advice to young researchers. If you want to classify organisms, you must go at the job one step at a time, starting with the single, isolated organ.

> First the variations in form of this organ should be observed, including paleontological evidence if possible; then its function or functions should be determined. With this knowledge endeavour to determine what was the primitive form of the organ and the various ways in which this primitive form has been modified, keeping in mind the relation of the changes in form of the organ to its functions. In other words, endeavour to read the action of natural selection upon the group of organisms as it is recorded in a single organ. The data thus obtained will aid in making a *provisional* classification of the group. (Comstock 1893, 41)

This does not read like the thinking of a man who has turned his back on natural selection as a causal tool of modern professional evolutionary biology.

Darwin as Popular Science

All of this said, the fact remains that for the first fifty years after the *Origin*, as a key to inquiry and explanation, natural selection was not a major success as a means to further professional science. More precisely, it was not a major success in those areas of science like paleontology that had no real need of it. It was a major success in those areas of science like entomology that had real need of it. Insects are obvious models for selection explanations because they are fast breeding and tend to have adaptations easily discoverable. If an insect looks like a twig, you can bet it is meant to look like a twig. With respect to the status of natural selection, the trouble was that anatomy and related subjects had high status as professional sciences. Entomology, often (as with Albert Brydges Farn, a civil servant) was the pursuit of amateurs, for whom the study of nature was more a hobby than a full-time profession. Even the more professional men tended not to have full-time jobs as scientists. Bates, for instance, thanks to the very welcome help of Darwin, became the assistant secretary to the Royal Geographical Society. A job, but something his friend Wallace was to describe (in his obituary) as a lifetime of drudgery. And agriculture? Well, everyone knows that its place in the academic pecking order is little above education.

This, however, is but one part of the story. From the beginning, everyone—the general public—knew of the *Origin*, knew of natural selection (and later sexual selection), realized its implications, and read about and discussed it non-stop. There really was a Darwinian Revolution, but it was mainly at the level of what we might call "popular science." The novelist Charles Dickens published a weekly magazine, *All the Year Round*, that over the years carried many well-known novels, including Dickens's own *Great Expectations*; Wilkie Collins's thriller *The Woman in White*; and Anthony Trollope's final Palliser novel, *The Duke's Children*. Its circulation was around one hundred thousand copies, which meant it was read by upwards of half a million good, respectable Victorians—wives and daughters, as well as the menfolk. Often, as in the Darwin family, in this pre-television age the whole group would gather on a regular basis for one member to read aloud to the others.

Darwin and natural selection are right there in this magazine in 1860, and in a way far more sympathetic than you might expect from the Huxley-Wilberforce story. The author, a geologist David Thomas Anstead, could hardly have been more positive.

> How, asks Mr. Darwin, . . . have all these exquisite adaptations of one part of the organisation to another part, and to the conditions of life, and of one distinct organic being to another, been perfected? He answers, they are so perfected by what he terms Natural Selection—the better chance which a better organised creature has of surviving its fellows—so termed in order to mark its relation to Man's power of selection. Man, by selection in the breeds of his domestic animals and the seedlings of his horticultural productions, can certainly effect great results, and can adapt organic beings to his own uses, through the accumulation of slight but useful variations given to him by the hand of Nature. But Natural Selection is a power incessantly ready for action, and is as immeasurably superior to man's feeble efforts, as the works of Nature are to those of Art. Natural Selection, therefore, according to Mr. Darwin—not independent creations—is the method through which

the Author of Nature has elaborated the providential fitness of His works to themselves and to all surrounding circumstances. ([Anstead] Anon. 1860a)

It is true that a lot of this is speculation, but take it seriously. "We are no longer to look at an organic being as a savage looks at a ship— as at something wholly beyond his comprehension; we are to regard every production of nature as one which has had a history; we are to contemplate every complex structure and instinct as the summing up of many contrivances, each useful to the possessor, nearly in the same way as when we look at any great mechanical invention as the summing up of the labour, the experience, the reason, and even the blunders, of numerous workmen" (1860b, 299). The worth of Darwin as a scientist and his moral standing as a human being are stressed and opponents are characterized as "timid."

There is similar material elsewhere, for instance, in a rival literary magazine, *The Cornhill Magazine*, edited by the novelist William M. Thackeray. It too introduced some of the great works of Victorian fiction, for instance, Thomas Hardy's *Far from the Madding Crowd.* It too had a one hundred thousand circulation, and it too took up the cause of natural selection. In 1862, it carried a little fiction about a visit to the zoo, where one sees in action Madame Natural Selection and her not-very-nice son Struggle for Existence. The wolf food supply is cut off. "As the pangs of hunger became sharper and sharper, the ravenous brutes set to devouring each other, the vigorous destroying the old, the healthy tearing the feeble limb from limb, till none were left but a single pair, male and female, the gauntest, savagest, and most powerful of all that savage group" (Dixon 1862, 317). Get to it now, says the dreadful child. Produce children as "wolfish" as you. Unfeeling perhaps, but even if wrong to be taken very seriously. "Still the book has given me more comprehensive views than I had before . . . Here we are offered a rational and a logical explanation of many things which hitherto have been explained very unsatisfactorily, if at all. It is conscientiously reasoned and has been patiently written. If it be not the truth, I cannot help respecting it as sincere effort after truth" (318).

It's all a question of where you look. In general Victorian culture, evolution is a given and Darwinian natural selection—sexual selection even more so—is known, talked about, used, and always there. You cannot read the great novels without seeing Darwinian themes at work. Often we are told explicitly that it is Darwin who is behind all of this. In George Gissing's *New Grub Street*, the journeyman writer who has guts and determination is the one who wins out and the more talented novelist, unable to compromise or fight for his ideas, loses and dies—leaving his wife to the survivor.

Not that she is about to complain—"though she had never opened one of Darwin's books, her knowledge of his main theories and illustrations was respectable" (Gissing 1891, 397). In poetry too. Young people, particularly young women, loved Darwin. This poem, by Constance Naden, is titled "Natural selection." Truly, it is sexual selection at work here.

I HAD found out a gift for my fair,
I had found where the cave men were laid:
Skulls, femur and pelvis were there,
And spears that of silex they made.

But he ne'er could be true, she averred,
Who would dig up an ancestor's grave—
And I loved her the more when I heard
Such foolish regard for the cave.

My shelves they are furnished with stones,
All sorted and labelled with care;
And a splendid collection of bones,
Each one of them ancient and rare;

One would think she might like to retire
To my study— she calls it a "hole"!
Not a fossil I heard her admire
But I begged it, or borrowed, or stole.

But there comes an idealess lad,
With a strut and a stare and a smirk;
And I watch, scientific, though sad,
The Law of Selection at work.

Of Science he had not a trace,
He seeks not the How and the Why,
But he sings with an amateur's grace,
And he dances much better than I.

And we know the more dandified males
By dance and by song win their wives—
'Tis a law that with *avis* prevails,
And ever in *Homo* survives.

Shall I rage as they whirl in the valse?
Shall I sneer as they carol and coo?

Ah no! for since Chloe is false
I'm certain that Darwin is true.

(Naden 1999, 207–208)

All a good joke, but serious for all that.

So grant that Darwin's thinking—Darwinian selection—entered the realm of popular science. Grant also, for it hardly needs stressing or arguing, that whatever happened to Darwinian thinking in the realm of professional science, this popular side to Darwinian thinking has persisted down to the present. Think of the dazzling series of essays by Stephen Jay Gould, published every month in *Natural History*, and then collected in such bestselling volumes as *Ever Since Darwin* (1977). Likewise the even-more phenomenal success of Richard Dawkins with the *Selfish Gene* (1976) and the *Blind Watchmaker* (1986). As a third, the stunning series of books by Edward O. Wilson, starting with his Pulitzer Prize-winning *On Human Nature* (1978). In the world of the general public, Darwinian thinking, meaning evolution through natural selection, was a huge success. Before Darwin, everyone knew about "nature red in tooth and claw." The phrase in fact comes from the most popular poem of the Victorian era, Alfred Tennyson's *In Memoriam* (1850). It was Darwin's genius to show that it was not just poetic fancy, but a central theme in the very serious world in which we all live. From this, Darwin had gone on to apply his ideas to all living things, most notably humans. Sex especially, a point much appreciated by those supposedly straight-laced Victorians. Choosing mates for Darwinian reasons is there through George Eliot and Henry James, on to Edith Wharton and D. H. Lawrence. It is even to be found in *Tarzan of the Apes*, as Jane ponders whether to go with the he-man Tarzan or the respectable and safe fellow who mistakenly thinks that he is the true Lord Greystoke. "Jane was not coldly calculating by nature, but training, environment and heredity had all combined to teach her to reason even in matters of the heart" (Burroughs 1912, 340). She makes the wrong choice, but fortunately Darwin will out, for this is rectified in the next of an unending series of sequels, the reading of which one saves for eternity.

Darwinian evolution at the popular level. The question now is about jacking up the ante as it were. Can we go on to speak of Darwinism as religion?

2 | Darwinism as Religion

HOW ARE WE TO ANSWER this question? First, obviously we want to know something about what we mean by religion. Here's the rub! As soon as we start asking that question, looking for necessary and sufficient conditions—a right-angled triangle is a triangle with one angle at 90^0—we run into problems. A religion involves a belief in a god or gods. What about the Unitarians, let alone the Buddhists? A religion involves a creed, a statement of what one is expected to believe. What about the Quakers? A religion has a priesthood. Quakers again, not to mention people like the Mormons who go rather more overboard about priesthoods than more staid Christians like Catholics think altogether decent. What about life after death? Although traditional Judaism made much about the survival of the Jews as a race, there wasn't much about the fate of individuals. Buddhists famously are working toward some kind of total nonbeing. So it goes. Most who think about these issues prefer to go more with some kind of cluster or polythetic definition—you list a number of pertinent features and possession of a number of these qualifies you as a religion. They don't have to be the same features for different religions or different branches of what is generally taken to be one religion.

Christianity

Let's cut down the scope of our question. Although there is no one feature or features that uniquely picks out a religion from a non-religion—Jehovah's Witnesses versus the Shriners, for instance—if we start to focus in on a particular religion, like Christianity or Islam, surely there are going to be claims or beliefs that are shared and recognized more or less by all within that religion? Perhaps not all of the claims all of the time, but enough to be defining and against which even exceptions can be handled and understood? Take Christianity. Overall, there is no question about Christianity being a religion. It is paradigmatically what we mean by a religion—God, creeds, priests, and so forth, going on through meaning to promises about the future. What about a core set of beliefs? To be a Christian you really have to accept a creator God—a powerful and loving God—and Jesus as His son. There is lots of discussion about what it means to be the son of God, but rest assured that Jesus is not just one of the chaps, a nice fellow but really no different from the rest of us. Jesus is divine and in some sense an aspect of the Creator God—he is not a second god in the Greek or Roman mode—and his coming to earth and living among us had—has—a purpose. For some reason or another, for all that we humans are "made in the image of God," we are not perfect. We are fallen and do wrong. Appreciating that we shall have to spend some time elaborating, the usual way of putting this is to say that we are tainted with "original sin." As such, we are in need of help in order to achieve the end that Christianity promises, salvation, meaning an eternity of bliss with God. Usually the death on the Cross and the subsequent resurrection are taken to be fundamental here, although, as is well known, there have been two thousand years of frenzied thinking by Christian theologians as to the true meaning of all of this. The point is that we could not do it on our own. We had to have God's help.

So, here's the bottom line for Christians, a line we can accept for the purposes of our discussion, notwithstanding that we must recognize that there are hundreds of variations on the theme. God exists and is good; Jesus is the son of God; we are the favored creation of God who cares about our well-being; we are tainted by or inclined to sin; unaided we cannot lift ourselves out of our situation; Jesus came and made the ultimate sacrifice to make possible our salvation. As we learn in the magnificent hymn by the Congregationalist minister Isaac Watts, published in 1707:

> When I survey the wondrous cross
> On which the Prince of glory died,
> My richest gain I count but loss,
> And pour contempt on all my pride.

Humans are sinners and we depend on God's help. Providence.

Darwin's Debt to the Church of England

Turn now to what I am calling Darwinism, meaning Darwinian thinking as a religion or religious perspective. Start with an important fact that in taking Christianity as my comparison religion—my mirror for the rest of this book—I am not being unduly ethnocentric or whatever. Thanks to Christianity being the parent of Darwinian thinking, Christianity has a privileged position in this discussion. Actually, and this is my justification for thus constraining the discussion in this book and from now on taking it as a given, Western Christianity has a privileged position in this discussion. Darwinism is Western Christianity's bastard offspring. As I noted sardonically about Darwin's views on women in the *Descent*, he may have been a great revolutionary. He was not a rebel. He made a new picture, but he made it from parts given to him in his background and training (Richards and Ruse 2016). We have seen this in his use of the *vera causa* principle and his very orthodox use of the thinking of Herschel and Whewell. We see this now as we plunge into the content of the *Origin*. All of the main moves come right out of the Church of England!

Go through the theory of the *Origin*, devised by a young man who had been raised a conventional Anglican—probably, thanks to his family's detestation of slavery, toward the evangelical end of the spectrum (Desmond and Moore 2009). A young man who, before he was sidetracked into a lifetime as a naturalist, thinking that the life of an Anglican parson might be the perfect occupation for someone of private means, went to the University of Cambridge, a Church of England institution where many of the teachers and professors were ordained priests. In seeking a Herschelian *vera causa*, Darwin was accepting explicitly that not only was he going with a force that is something akin to human will force—it is we who are pulling the stone down back through the string—but he was working with a force that ultimately is God's force. This is not pantheism, but belief that in some sense God is present in His creation. At the time of writing and publishing the *Origin*, this was very much Darwin's theology. Deism left much untouched and in this respect is very Christian—"he is before all things and by him all things exist" (Colossians 1:17). It found its way right into the text.

> Authors of the highest eminence seem to be fully satisfied with the view that each species has been independently created. To my mind it accords better with what we know of the laws impressed on matter by the Creator, that the

> production and extinction of the past and present inhabitants of the world should have been due to secondary causes, like those determining the birth and death of the individual. When I view all beings not as special creations, but as the lineal descendants of some few beings which lived long before the first bed of the Silurian system was deposited, they seem to me to become ennobled. (Darwin 1859, 488–489)

This is not the world of Cartesian *res extensa*, just inert matter in motion. It is a sentiment that Darwin left unchanged through the six editions of the *Origin*, to the last of 1872, by which time his deism had changed or faded into a form of agnosticism. Darwin's God kept working away, and as to His existence? Well, Darwin left that as an exercise for the reader.

Look at the key moves in Darwin's theorizing. First to the struggle for existence. Explicitly Darwin (1859) says of this conflict that it "is the doctrine of Malthus applied with manifold force to the whole animal and vegetable kingdoms" (63). Darwin refers to the essay on population by the Reverend Thomas Robert Malthus (1798; sixth edition 1826). This divine was an Anglican clergyman who, writing at the end of the eighteenth century, worried that populations potentially go up geometrically in numbers, whereas food supplies can never increase more than at an arithmetic rate. There are bound to be clashes between people, what Malthus himself called "struggles for existence." Rather than (as is traditional) seeing Malthus's essay simply as the callous musings of a middle-class cleric laying down the law for the unwashed masses, it is now thought more discerning to recognize that Malthus is offering an exercise in natural theology (Mayhew 2014). He, like many of his fellow countrymen at the end of the eighteenth century, was certainly worried about the huge population increases that were occurring in cities all over Britain. His was an attempt to put it all in context, in God's context. The clash between population growth and limited resources was God's way of making us humans get up off our backs and set to work. If everything were provided, no matter what we did, then we would do nothing. God is stern but has our welfare at heart.

Second, we have natural selection and adaptation. The problem of adaptation is the central problem of the empirical approach to God. It is highlighted and discussed by Archdeacon William Paley in his textbook, *Natural Theology* (1802), a work Darwin said, humorously but truly, he could have virtually written down from memory. The hand and the eye could not have come about by chance. Random laws in motion do not produce intricate functioning characteristics. They do not produce adaptation.

There has to be a cause. This for Paley was God. This for Darwin was God, but a God working at a distance through blind unbroken natural law. People like Huxley had trouble with selection precisely because adaptation was not his problem.

Of course, it wasn't just Christians who seized on the problem of adaptation—it is right there in Plato's *Phaedo* (Ruse 2017b). No problem. All the great thinkers—starting with God Himself—have been Englishmen manqués. Paley and Plato were both very much part of the comfortable, establishment religion of Darwin's youth. The world is adapted. Why? Many a clergyman was happy to give an answer. God! Secure in a comfortable living, these Anglican parsons often had a near-professional interest in biology, justifying it all by arguing that ferreting out adaptation was, just as much as ministering to the poor, doing God's work. At Cambridge, Darwin became an avid beetle collector and (hardly surprising since he was planning on becoming a minister himself) got this party line all of the time. The definitive work on British insects was the *Introduction to Entomology* (1815–1828) by the Reverend William Kirby and William Spence. Explicit is the message that looking at and collecting insects is no purely secular activity. For good Anglicans, it is as being in church: "no study affords a fairer opportunity of leading the young mind by a natural and pleasing path to the great truths of Religion, and of impressing it with the liveliest ideas of the power, wisdom, and goodness of the Creator" (xvi).

> There is a book, who runs may read,
> which heavenly truth imparts,
> and all the lore its scholars need,
> pure eyes and Christian hearts.
>
> The works of God above, below,
> within us and around,
> are pages in that book, to shew
> how God himself is found.

A hymn written by the Anglican John Keble in 1819.

Moving on to the fact of evolution, the division of labor was crucial in the explanation of speciation. It hardly needs saying how theological a doctrine this was, with the "Invisible Hand" having devised it as a method of efficient work and standing behind its operation, seeing that the end results are good. No less theological is the "tree of life."

> [22] And the LORD God said, Behold, the man is become as one of us, to know good and evil: and now, lest he put forth his hand, and take also of the tree of life, and eat, and live for ever:
>
> [23] Therefore the LORD God sent him forth from the garden of Eden, to till the ground from whence he was taken.
>
> [24] So he drove out the man; and he placed at the east of the garden of Eden Cherubims, and a flaming sword which turned every way, to keep the way of the tree of life. (Genesis 3)

Let me make clear what I am claiming at this point and what I am not claiming. I am not claiming that because Darwin's theory was conceived in Anglican theology, that means it is a religious theory. Contrary to Freud, it is possible to move on from your origins and childhood. I am my mother's son, but I am not female. I do however share with her the fact of being human, with overlapping abilities and interests and desires, starting with speaking the same language and smothering everything we eat with brown sauce. That is the relationship I see between Darwinians and Anglicans. I have no reason to think that a modern-day professional evolutionary biologist is doing natural theology. I don't even think that writing for the popular market about Darwin is necessarily doing natural theology. What I would say is that the Anglican origins prepare the way for a religious interpretation or conversion. In Darwinian theory, we are not talking of something of a totally different logical type from Christianity—we are not, for instance, trying to compare Christianity to the Beatles, for all that John Lennon said they were more popular than Jesus. More than this, we are preparing the way for Darwinians to think about the big issues—God, humans, morality, sin—drawing from a shared background with Christians. The solutions are not necessarily going to be the same, but the problems and questions are shared.

God

Move the discussion along. Accept the origins of Darwinism. Accept also that because the origins were Christian, this does not necessarily imply that Darwinism is a religion. Start to dig into the content, beginning with the biggest issue of them all, God. For the Christian, God exists, is creator, and is all-powerful and all good. These are bottom-line demands. Does this conflict with Darwinism or show that the Darwinian answer is less scientific than religious? Some people, like Richard Dawkins (2006), think there is an out-and-out contradiction. Like those weather indicators, where when it is wet the little man comes out of the house and when dry the little woman—a function one presumes of the state of the catgut controlling them—if you are on-board with Christianity you are off with Darwinism, and if you are on-board with Darwinism you are off with Christianity. All of which is true, obviously, in some sense. One cannot accept a Darwinian perspective on time and continue with a literal reading of Genesis. One cannot accept a literal worldwide Noah's Flood. I very much doubt one is going to be all that comfortable with the parting of the Red Sea.

This is not the final word. Many (most) Christians would contest the claim that there really is opposition and refutation here. Genesis is not science. Perhaps the days are long periods of time. Perhaps there are unrecorded eons. Darwin doesn't clash with Christianity. The same with a grown-up reading of the story of the Deluge. Whether or not Christians are prepared to agree, however, is beside the point. Either way, you are hardly making Darwinism into a religion. True, but don't stop here. Press matters a little harder. Take the problem of natural evil, the misfortunes that come about through the regular workings of the laws of nature, the Lisbon earthquake or the life-threatening leukemia of the young child. Here surely, we do start to think in terms of alternatives, meaning alternatives from but in the same mode as traditional religious perspectives. Moreover, here we start to find an area that Darwinians have discussed at length in terms of their theory. In a well-known letter written (just after the *Origin*) to his American friend, the Harvard botanist Asa Gray, Darwin opined:

> With respect to the theological view of the question; this is always painful to me. — I am bewildered. — I had no intention to write atheistically. But I own that I cannot see, as plainly as others do, & as I shd. wish to do, evidence of design & beneficence on all sides of us. There seems to me too much misery in the world. I cannot persuade myself that a beneficent & omnipotent God would have designedly created the Ichneumonidae with the express intention of their feeding within the living bodies of caterpillars, or that a cat should play with mice. Not believing this, I see no necessity in the belief that the eye was expressly designed. (Darwin 1985–, 8, 224)

It is a worry that continues down to the present. Richard Dawkins, having quoted Darwin, gnaws away at the issue. Cheetahs seem wonderfully designed to kill antelopes. "The teeth, claws, eyes, nose, leg muscles, backbone and brain of a cheetah are all precisely what we should expect if God's purpose in designing cheetahs was to maximize deaths among antelopes" (Dawkins 1995, 105). Conversely, "we find equally impressive evidence of design for precisely the opposite end: the survival of antelopes and starvation among cheetahs." One could almost imagine that we have two gods, making the different animals, and then competing. If there is indeed but one god who made both animals, then what is going on? What sort of god makes this sort of encounter? "Is He a sadist who enjoys spectator blood sports? Is He trying to avoid overpopulation in the mammals of Africa? Is He maneuvering to maximize David Attenborough's television ratings?"

Here we do seem to be getting closer to the question of the status of Darwinism. Although, note that now we are starting to shift the domain of discussion. All of this animal suffering isn't a problem in the realm of Darwinian theory as regular science. A lion catches and kills an antelope. Too bad! Our questions here are focused on the empirical. Why is it always the females who do the hunting? Why do the females hunt in pairs or groups? Who is related to whom? Is it more commonly sisters who are joined up? Our questions here are not theological but scientific; however, already in the passages quoted above, we are no longer "here." Darwin and Dawkins have moved on to the meaning of it all. The meaning of it all in what can only be described as a religious meaning. Not the question: Does the fighting disprove or challenge Christianity? Rather the question: Does the fighting as something cruel and painful, central to and highlighted by Darwinism, disprove Christianity? As it happens, Darwin and Dawkins differed in their answers to this question. Dawkins thinks it points to the end of God. Darwin is more nuanced. Continuing his letter to Gray:

> On the other hand I cannot anyhow be contented to view this wonderful universe & especially the nature of man, & to conclude that everything is the result of brute force. I am inclined to look at everything as resulting from designed laws, with the details, whether good or bad, left to the working out of what we may call chance.

The point is here however not really whether the Darwinian take on pain and suffering is enough to refute the existence of the Christian God. I myself am inclined to think not (Ruse 2001). The point is that Darwinians do worry about this in theological terms—pain and suffering is not just pain and suffering but is evil—and that many take it to be a refutation of God's existence. We go from the science of the *Selfish Gene*—the greatest popular (and very pro) work on Darwinian theory—to the theology of the *God Delusion*—the greatest popular (and very pro) work on atheism. If this isn't getting us into the realm of religion, I don't know what is. "The God of the Old Testament is arguably the most unpleasant character in all fiction: jealous and proud of it; a petty, unjust, unforgiving control-freak;

a vindictive, bloodthirsty ethnic cleanser; a misogynistic, homophobic, racist, infanticidal, genocidal, filicidal, pestilential, megalomaniacal, sadomasochistic, capriciously malevolent bully" (Dawkins 2006, 1). This is not the voice of Popperian objective science. It is the voice of an Old Testament prophet.

Actually, Darwinism (as I am calling it) does not strictly demand the nonexistence of God. It couldn't really because there are all sorts of questions about ultimate causes and natures—Could God be a necessary being?—to which science speaks not at all, let alone in refutation (Ruse 2001, 2010). When Darwin did become a nonbeliever, it had little directly to do with his science. It is true that, in the *Descent*, he gave a naturalistic account of religion's origins. Faith-based systems are dismissed in a Humean fashion as all one big mistake: "my dog, a full-grown and very sensible animal, was lying on the lawn during a hot and still day; but at a little distance a slight breeze occasionally moved an open parasol, which would have been wholly disregarded by the dog, had any one stood near it. As it was, every time that the parasol slightly moved, the dog growled fiercely and barked" (Darwin 1871, 1, 67). Instinct kicked in. "He must, I think, have reasoned to himself in a rapid and unconscious manner, that movement without any apparent cause indicated the presence of some strange living agent, and no stranger had a right to be on his territory." But, as Darwin then went on to say, this kind of argument is only definitive for those who already don't believe. The real reason for Darwin's agnosticism was that he could not accept the theology of Christianity, that morally upright nonbelievers like his own father were condemned to everlasting damnation simply because of their nonbelief (Darwin 1958).

Whether or not Darwin was right about this, whether or not Dawkins is right about his reading of the situation, the main message that people took from the *Origin* was somewhat broader, covering the range of people from Darwin to Dawkins. It had less to do with God's actual existence or not. It was rather that He is indifferent. He doesn't send Jesus down to earth for our salvation. He couldn't care less. That is what hurts and what drives Darwinism as religion. Listen to the poet Thomas Hardy—raised a good Anglican—but now, after reading the *Origin*, cast into the darkness of despair. A world without meaning. "Hap," written in 1866:

> IF but some vengeful god would call to me
> From up the sky, and laugh: "Thou suffering thing,
> Know that thy sorrow is my ecstasy,
> That thy love's loss is my hate's profiting!"

Then would I bear, and clench myself, and die,
Steeled by the sense of ire unmerited;
Half-eased, too, that a Powerfuller than I
Had willed and meted me the tears I shed.

But not so. How arrives it joy lies slain,
And why unblooms the best hope ever sown?
— Crass Casualty obstructs the sun and rain,
And dicing Time for gladness casts a moan. . . .
These purblind Doomsters had as readily strown
Blisses about my pilgrimage as pain.

At heart, God or no God, Richard Dawkins feels exactly the same way:

> In a universe of blind physical forces and genetic replication, some people are going to get hurt, other people are going to get lucky, and you won't find any rhyme or reason in it, nor any justice. The universe we observe has precisely the properties we should expect if there is, at bottom, no design, no purpose, no evil and no good, nothing but blind, pitiless indifference. (Dawkins 1995, 133)

Humans

Actually, one never gets from Dawkins the despair one finds in Hardy. Like all good trusting evangelicals, our modern writer has been plucked forth from the Slough of Despond. "Give me thy hand: so he gave him his hand, and he drew him out, and set him upon sound ground, and bid him go on his way." This all happens when we move on from God to the other factor in the equation, humankind. For the Christian, because God created us in His image, we have a special status in the Creation. To refer again to warthogs. You may love warthogs—actually, after a trip to Zimbabwe, I have a sneaking affection for them—but they are not God's chosen. He did not die on the Cross for their eternal salvation. What about Darwinism? In the happy world of Richard Dawkins, humans have as much superior status as they do in the world of the Christians, and warthogs I am afraid lose out yet again. For Dawkins, therefore, just as the Christian finds salvation from the blackness of sin through the blood of the lamb—Providence—the Darwinian finds salvation from the existential nothingness of an indifferent world through the process of evolution—progress. No original state, sinful or not, could hold us back. Humans are special.

Note again that we are not talking strict science here. Darwin's theory of evolution includes us humans in the picture; but, qua science, we see a break from the kinds of claims about human superiority that religious people want to make. No one says that we humans are just the same as other organisms. Students of human evolution, "paleoanthropologists," spend their careers showing how and why we are different. However, in the world of science, ours is not a different "different." We are different in the way that all species are different. We have our adaptations. They have their adaptations. We can reason. Birds can fly. Fish can swim. Ultimately, we are part of the tree of life produced by natural selection. Contra to a Christian perspective, we are not special. Indeed, Darwin's theory itself gives us a reason why we are not special. Darwinian evolutionary theory is non-progressive. On the one hand, as we have seen stressed, the new variations—the building blocks of evolution—are random, in the sense of not occurring according to need. Darwin argued this at length with Asa Gray who was convinced that the variations have direction. Gray wrote:

> But there is room only for the general declaration that we cannot think the Cosmos a series which began with chaos and ends with mind, or of which mind is a result: that, if, by the successive origination of species and organs through natural agencies, the author means a series of events which succeed each other irrespective of a continued directing intelligence—events which mind does not order and shape to destined ends—then he has not established that doctrine, nor advanced toward its establishment, but has accumulated improbabilities beyond all belief. (Gray 1860, 183)

Darwin would have nothing of this. Picking up on a metaphor used by Gray about water being channeled down certain streams, he wrote, "If we assume that each particular variation was from the beginning of all time preordained, the plasticity of organisation, which leads to many injurious deviations of structure, as well as that redundant power of reproduction which inevitably leads to a struggle for existence, and, as a consequence, to the natural selection or survival of the fittest, must appear to us superfluous laws of nature" (Darwin 1868, 2, 428). Darwin knew the score.

On the other hand, compounding the problem, or rather making even more definitive the case against progress, natural selection is relativistic. What succeeds is what succeeds. This is not a tautology. There is the empirical expectation that what happens in one place and time will hold in other places and times. What it is saying is that there is no ultimate better or worse. Against the ice, a white coat is better camouflage. Against the tundra, a darker coat is better. If global warming continues, white coats will never again be better. Darwin saw this. In a notebook Darwin kept in the late 1830s, just when he was discovering his theory, he wrote: "there is no NECESSARY tendency in the simple animals to become complicated" (Barrett et al. 1987, E 95). Then, ten years later, on the flyleaf of his copy of *Vestiges*, Darwin cautioned himself never to use the terms "higher" or "lower." A sentiment shared by Darwinian evolutionists down to the present. A year or two back, the eminent paleontologist Jack Sepkoski commented colorfully. "I see intelligence as just one of a variety of adaptations among tetrapods for survival. Running fast in a herd while being as dumb as shit, I think, is a very good adaptation for survival" (Ruse 1996, 486). Cow power rules supreme!

So much for the science. If you are talking human superiority, science is of little help. It is hard to exaggerate this point. Darwinian science is about as helpful as your average airline, lost-and-found-luggage-claim desk. What Darwin knew, as everyone knew, no one was going to stop with the science. Pre-Darwinian evolutionists sounded the trumpet, for they were progressionists to a person. Monad to man. Thus Erasmus Darwin:

> Organic Life beneath the shoreless waves
> Was born and nurs'd in Ocean's pearly caves;
> First forms minute, unseen by spheric glass,
> Move on the mud, or pierce the watery mass;
> These, as successive generations bloom,
> New powers acquire, and larger limbs assume;
> Whence countless groups of vegetation spring,
> And breathing realms of fin, and feet, and wing.
>
> Thus the tall Oak, the giant of the wood,
> Which bears Britannia's thunders on the flood;
> The Whale, unmeasured monster of the main,
> The lordly Lion, monarch of the plain,
> The Eagle soaring in the realms of air,
> Whose eye undazzled drinks the solar glare,
> Imperious man, who rules the bestial crowd,
> Of language, reason, and reflection proud,
> With brow erect who scorns this earthy sod,
> And styles himself the image of his God;
> Arose from rudiments of form and sense,
> An embryon point, or microscopic ens!
>
> (Darwin 1803, 1, 11, 295–314)

Fifty years later, just before the *Origin*, exactly the same message was being preached. Darwin's fellow Englishman and evolutionist, Herbert Spencer, wrote:

> Now we propose in the first place to show, that this law of organic progress is the law of all progress. Whether it be in the development of the Earth, in the development of Life upon its surface, in the development of Society, of Government, of Manufactures, of Commerce, of Language, Literature, Science, Art, this same evolution of the simple into the complex, through successive differentiations, hold throughout. From the earliest traceable cosmical changes down to the latest results of civilization, we shall find that the transformation of the homogeneous into the heterogeneous is that in which Progress essentially consists. (Spencer 1857, 2–3)

All of these people thought that the progress was built-in in some way.

So what happened after the *Origin* appeared? Straight off, Darwinians (starting with Darwin) went with the tide. No one, no self-respecting Victorian, was going to take seriously the idea that we humans—we Europeans—are still down at the warthog level or below. Darwinians built in the thoroughly value-impregnated belief that evolution—evolution through natural selection—leads progressively up to humankind. This is Darwin himself, twenty years before the *Origin*, just before he discovered natural selection.

> We see gradation to mans mind in Vertebrate Kindgdom in more instincts in rodents than in other animals & again in mans mind, in different races. being unequally developed.
>
> ? is not Elephant intellectually developed amongst Pachydermata like man amongst Monkeys or dogs in Carnivora. —
>
> Man in his arrogance thinks himself a great work worthy the interposition of a deity, more humble & I believe truer to consider him created from animals. (C, 196)

There is progress to humans although notice how Darwin is putting things in an evolutionary context and describing his position not as superior to another evolutionary position, Lamarck for instance, but to the Christian position. This is a man thinking in a theological mode.

After the discovery of selection, there is still progress albeit with an increasing recognition that selection changes the rules of the game. You could not now get a teleological upward direction even if you wanted it.

> The enormous number of animals in the world depends of their varied structure & complexity. — hence as the forms became complicated, they opened fresh means of adding to their complexity. — but yet there is no necessary tendency in the simple animals to become complicated although all perhaps will have done so from the new relations caused by the advancing complexity of others. — It may be said, why should there not be at any time as many species tending to dis-developement (some probably always have done so, as the simplest fish), my answer is because, if we begin with the simplest forms & suppose them to have changed, their very changes ~~ton~~ tend to give rise to others. (E 95–97)

Although the *Origin* does not discuss humans, it makes clear Darwin's belief in progress. "The inhabitants of each successive period in the world's history have beaten their predecessors in the race for life, and are, in so far, higher in the scale of nature; and this may account for that vague yet ill-defined sentiment, felt by many palæontologists, that organisation on the whole has progressed" (Darwin 1859, 345). The closing passage to the book repeats that sentiment. The question is how selection brings this about. Through competition, obviously, but precisely how? By the third edition of the *Origin*, 1861, Darwin was confident that he had the answer. He invoked what today's Darwinians call "arms races"—lines compete against each other, and one line gets better and better. Eventually this comparative improvement is translated into some form of absolute improvement, *Homo sapiens*.

> If we look at the differentiation and specialisation of the several organs of each being when adult (and this will include the advancement of the brain for intellectual purposes) as the best standard of highness of organisation, natural selection clearly leads towards highness; for all physiologists admit that the specialisation of organs, inasmuch as they perform in this state their functions better, is an advantage to each being; and hence the accumulation of variations tending towards specialisation is within the scope of natural selection. (Darwin 1861, 134)

Natural selection cannot guarantee anything, but everything is probably going to be just fine. The great success of capitalism and British industry leads the way.

The Persistence of Progress

This kind of thinking persists in Darwinian circles right down to the present. Richard Dawkins again. "Directionalist common sense surely wins on the very long time scale: once there was only blue-green slime and now there are sharp-eyed metazoa" (Dawkins and Krebs 1979, 508). This is all thanks to arms races.

> Notwithstanding [Stephen Jay] Gould's just skepticism over the tendency to label each era by its newest arrivals, there really is a good possibility that major innovations in embryological technique open up new vistas of evolutionary possibility and that these constitute genuinely progressive improvements. . . . The origin of the chromosome, of the bounded cell, of organized meiosis, diploidy and sex, of the eucaryotic cell, of multicellularity, of gastrulation, of molluscan torsion, of segmentation – each of these may have constituted a watershed event in the history of life. Not just in the normal Darwinian sense of assisting individuals to survive and reproduce, but watershed in the sense of boosting evolution itself in ways that seem entitled to the label progressive. It may well be that after, say, the invention of multicellularity, or the invention of metamerism, evolution was never the same again. In this sense, there may be a one-way ratchet of progressive innovation in evolution. (Dawkins 1997, 1019–1020)

Explicitly Dawkins argues that evolution is cumulative. Today's arms races are ever increasingly electronic as both sides build bigger and better computers. So too in the animal world. Referring to something called the Encephalization Quotient, a kind of cross-species IQ equivalent, Dawkins writes: "The fact that humans have an EQ of 7 and hippos an EQ of 0.3 may not literally mean that humans are 23 times as clever as hippos! But the EQ as measured is probably telling us *something* about how much 'computing power' an animal probably has in its head, over and above the irreducible amount of computing power needed for the routine running of its large or small body" (Dawkins 1986, 189). No prizes for guessing what that something is.

There are today other suggestions for generating progress to humans from a Darwinian perspective. Simon Conway Morris, in an argument that was once floated by Stephen Jay Gould (1985) of all people—as Dawkins hints, Gould was notoriously skeptical about progress—argues that there are ecological niches, existing independently of life, that organisms seek out and occupy. Conway Morris instances the saber-tooth tiger niche, a niche for a cat-like mammal with teeth that sheer and stab rather than bite. It was occupied independently by placental mammals and by marsupials. He suggests: "If brains can get big independently and provide a neural machine capable of handling a highly complex environment, then perhaps there are other parallels, other convergences that drive some groups towards complexity" (Conway Morris 2003, 196). Continuing: "We may be unique, but paradoxically those properties that define our uniqueness can still be inherent in the evolutionary process. In other words, if we humans had not evolved then something more-or-less identical would have emerged sooner or later."

Really more Spencerian (as they acknowledge) than Darwinian, Duke University colleagues biologist Daniel McShea and philosopher Robert Brandon have come up with—a term for production that covers "invents" as well as "discovers"—what they proudly call "biology's first law." Named the "zero-force evolutionary law," or ZFEL, its general formulation runs: "In any evolutionary system in which there is variation and heredity, there is a tendency for diversity and complexity to increase, one that is always present but that may be opposed or augmented by natural selection, other forces, or constraints acting on diversity or complexity" (McShea and Brandon 2010, 57). It holds, apparently, the status in evolutionary biology of Newton's first law of motion—a background condition of stability. Although the authors are (as tends to be the case at these sorts of times) fairly easygoing on their understandings of complexity and diversity—number of parts, number of kinds—the claim is that the natural tendency to complexity—parts tend to be added on—generates new organic variations and hence types. Somewhat typical for these sorts of discussions, it is not always obvious whether the claim is that adaptation is created in this way or if adaptation is now irrelevant. Probably more the former: "We raise the possibility that complex adaptive structures arise spontaneously in organisms with excess part types. One could call this self-organization. But it is more accurately described as the consequence of the explosion of combinatorial possibilities that naturally accompanies the interaction of a large diversity of arbitrary part types" (58).

I am not offering these progress-generating suggestions as exemplars of good, professional science. Certainly, not good, professional, Darwinian science. The whole point of the offering is to deny this. On empirical grounds, there are many critics of arms races. They do happen, but nothing like as often or inevitably as enthusiasts suggest (Ruse 1996). There is certainly no guarantee that they will lead to humans. People who push the computer analogy should really stop looking at their screens and get out into the fresh air. Turning to niches, they simply don't exist independently of their inhabitants. It is always a two-way business (Lewontin 2011). The distinctive insects who live in the zone at the tops of trees in the Brazilian rainforests did not find the niche. They created it and evolved to fit it in tandem. Again, even if there is a cultural niche, it might have been forever blocked to us. Almost certainly if the dinosaurs had gone on living. In any case, perhaps there are other cultural or "cultural" niches with conscious minds way better than ours, already somewhere in the universe occupied by far superior beings. As for ZFEL. It is strictly for people who believe in the tooth fairy. Blind laws don't lead to sophisticated, functioning complexity. At most, monkeys can type Shakespeare and undirected laws can create humans if you have an infinite number of multiverses and somewhere, sometime, it is bound to happen. But does one want to talk in terms of "progress" in such situations, with universes littered with garbled, unproducible dramas and billions of would-be philosophers with the IQs of turnips?

Progress to humans, perhaps. More wishing or faith than straight science. The claim certainly is central to Darwinism and it certainly is building in values. It is creating a religious world picture to challenge the religious world picture of the Christian. A point fully noticed by perceptive Christians. In his magisterial work, *The Nature and Destiny of Man*, based on the Gifford Lectures of 1939, the American, Lutheran theologian Reinhold Niebuhr, unsurprisingly given the looming threat of Nazi Germany, wrote strongly against the fallacy of progress. "The spiritual hatred and the lethal effectiveness of 'civilized' conflicts, compared with tribal warfare or battles in the animal world, are one of the many examples of the new evil which arises on a new level of maturity" (Niebuhr 1943, 315). Supporting claims that we are dealing with a secular metaphysics that has its roots in Christianity, Niebuhr writes, "The idea of progress is only possible on the ground of a Christian culture. It is a secularized version of Biblical apocalypse and the Hebraic sense of a meaningful history, in contrast to the meaningless history of the Greeks" (Niebuhr 1941, 24). Significantly, Niebuhr notes that this secularization crucially involved eliminating the Christian notion of sin. "But since the Christian doctrine of the sinfulness of man is eliminated, a complicating factor in the Christian philosophy is removed and the way is open to simple interpretations of history, which relate historical process as closely as possible to biological process . . ." Note how biology enters in here. "Darwinism is used to express the mode of historical optimism in the nineteenth century, and the biological idea of the survival of the fittest becomes the bearer of historical optimism" (Niebuhr 1943, 165). Niebuhr knew a rival religion when he saw one.

Morality

Finally ask about what the Christian picture expects of us humans. We have an obligation to do things to please God—to worship Him and to help our fellow humans. There is debate over whether helping others is a way to buy our tickets to heaven or more to show our gratitude to God for all that He does for us, but no matter here. We have strong moral obligations. It is the same for Darwinism. For Darwinism, not for Darwinian theory. In the *Descent*, Darwin gave an extended discussion of the evolution of morality, yet in this context made no prescriptions. What Darwin was in the business of showing was that cooperation is just as good an adaptation as fighting and brute competition. "It must not be forgotten that although a high standard of morality gives but a slight or no advantage to each individual man and his children over the other men of the same tribe, yet that an advancement in the standard of morality and an increase in the number of well-endowed men will certainly give an immense advantage to one tribe over another" (Darwin 1871, 1, 166). It is clear that Darwin thought tribes are groups of interrelated people or people who think they are interrelated. This means that we have a situation rather like the bees and other social insects. If I help others at my own expense, selection is happy with that so long as those others are relatives. Unlike others, Wallace as we shall see, Darwin never had much time for the group as such benefiting at the expense of the individual. In today's terminology, he was an individual selectionist, not a group selectionist.

Neither Darwin nor anyone else wanted to stop there. Immediately, he and his fellows were into prescriptions about the right way to run individual lives as well as societies generally. As one who benefited immensely from the Industrial Revolution, almost expectedly the *Descent* is laced with encomia about the virtues of capitalism. "In all civilised countries man accumulates property and bequeaths it to his children. So that the children in the same country do not by any means start fair in the race for success. But this is far from an unmixed evil; for without the accumulation of capital the arts could not progress; and it is chiefly through their power that the civilised races have extended, and are now everywhere extending, their range, so as to take the place of the lower races" (1, 169). Darwin did have the prudence to add that he thought this was fine by natural selection. Capitalism frees the superior from daily labor so they can devote their time and talents to the betterment of all humanity. "The presence of a body of well-instructed men, who have not to labour for their daily bread, is important to a degree which cannot be over-estimated; as all high intellectual work is carried on by them, and on such work material progress of all kinds mainly depends, not to mention other and higher advantages." This is a man who took Plato's *Republic* seriously.

On a day-by-day basis, morally speaking, Darwinism is in much the same business as Christianity. This comes as a surprise to many, who think that the nineteenth-century, pragmatist philosopher Charles Sanders Peirce hit the nail on the head when he said: "The *Origin of Species* of Darwin merely extends politico-economical views of progress to the entire realm of animal and vegetable life." Elaborating: "Among animals, the mere mechanical individualism is vastly re-enforced as a power making for good by the animal's ruthless greed. As Darwin puts it on his title-page, it is the struggle for existence; and he should have added for his motto: Every individual for himself, and the Devil take the hindmost! Jesus, in his sermon on the Mount, expressed a different opinion." Concluding that "the conviction of the nineteenth century is that progress takes place by virtue of every individual's striving for himself with all his might and trampling his neighbor under foot whenever he gets a chance to do so. This may accurately be called the Gospel of Greed" (Peirce 1893, 182).

I will leave the Christian side of things until the next chapter, where we shall see that things are slightly more complex than Peirce suggests, but the fact is that Darwinism is simply not the Gospel of Greed. The Darwinians always accepted fully the obligation to feed the hungry and to help the sick. The Darwin family worked closely with the local vicar helping the unfortunate in their society. Thomas Henry Huxley always argued for compulsory Bible study in state schools for the moral uplift. No rude bits from the psalms, however! This said, it is true that there is a subtle shift. For the Christian—not that you would know it when you consider some of the very greatest of Christian saints—meekness is a virtue. "Blessed *are* the meek: for they shall inherit the earth" (Matthew 5:5). Darwinians wanted none of that nonsense (Ruse 2017a). Life is a challenge, a struggle. The virtuous person is the one who gets up and fights. The one who shows courage, vim, and vigor. The classic example is Jack London's overheated story of dogs in the Far North.

> His teeth closed on Spitz's left fore leg. There was a crunch of breaking bone, and the white dog faced him on three legs. Thrice he tried to knock him over, then repeated the trick and broke the right fore leg. Despite the pain and helplessness, Spitz struggled madly to keep up. He saw the silent circle, with gleaming eyes, lolling tongues, and silvery breaths drifting upward, closing in upon him as he had seen similar circles close in upon beaten antagonists in the past. Only this time he was the one who was beaten.
>
> There was no hope for him. Buck was inexorable. Mercy was a thing reserved for gentler climes. (London [1903] 1990, 24)

Don't, however, read this as just a crude example of (what has come to be called) "Social Darwinism" in action. London is more aware than this. Buck doesn't just win for its own sake. He wins so he can improve things. No sooner does he become top dog than he turns into a high school principal. The other dogs are whipped into line, slackers and the inadequate are chastised and made to improve.

> Pike, who pulled at Buck's heels, and who never put an ounce more of his weight against the breast-band than he was compelled to do, was swiftly and repeatedly shaken for loafing; and ere the first day was done he was pulling more than ever before in his life. The first night in camp, Joe, the sour one, was punished roundly—a thing that Spitz had never succeeded in doing. Buck simply smothered him by virtue of superior weight, and cut him up till he ceased snapping and began to whine for mercy. (26)

Morality, but morality infused with the spirit of the age, or rather of the science. Thomas Henry Huxley is the paradigm of the new Darwinian man, for all that he had doubts about Darwinian selection as a cause (Desmond 1997). He worked himself into nervous breakdown after nervous breakdown as he pushed science, as he pushed evolution, as he reformed medical education, as he sat on royal commission after royal commission, as he penned one brilliant essay after another, as he delved into philosophy and wrote a book on Hume, as he took the then prime minister William Gladstone to task over the Gadarine Swine, as he smoked himself quite literally to death. At times, he irritates immensely as the worst kind of Victorian prig. Unlike the higher-class-level Darwins, who felt no such inhibitions, because the novelist was living with a man to whom she was not married, Huxley would not let his wife visit George Eliot. But then, you just have to stand back in admiration at the pure moral drive and vigor. He was truly a great man.

Religion?

Note again what I am arguing and what I am not arguing. I am not saying that Darwin's theory supports capitalism or urges you to become a school principal or even to feed the poor and succor the sick. Indeed, if you follow—as do I—the David Hume distinction between matters of fact and matters of moral value, you do not think that a science can even in principle do this at all. I am saying that this is the sort of thing that Darwinians, starting with Darwin himself, did from the first. I am also saying that once you get into values like this, you are getting into the domain of religion. And, if you embed your thinking and your actions in a general world picture, this looks very much like a religion, or at least a secular religious perspective.

It is important to note that Darwin and his supporters knew what they were about. Deliberately, Huxley split off his professional science from his popular science, but as deliberately made sure that each had its important place. A student, Father Hahn S. J., who studied with him in 1876, wrote: "One day when I was talking to him, our conversation turned upon evolution. 'There is one thing about you I cannot understand,' I said, 'and should like a word in explanation. For several months now I have been attending your course, and I have never heard you mention evolution, while in your public lectures everywhere you openly proclaim yourself an evolutionist'" (Huxley 1900, 428). Huxley's response was simple. Anatomy, embryology, physiology (which Huxley got assistants to teach) are part of professional science. It is for these that students come to be trained. As one who was as much an administrator as teacher, Huxley knew full well that in the end it is financial support that counts. He sold the medical profession—its confidence shattered because of its inadequacy during the Crimean War—on the idea that he and his fellows could teach future doctors the basic science, as part of a new, revised curriculum, and that they then could take over and teach the hands-on medicine. Evolutionary thinking simply does not cure a pain in the belly. Evolution, Darwinian evolution, has another function: to offer an alternative to the conventional Christian religion. It was this on which Huxley lectured—natural selection and all—night after night, in working man's club after working man's club, in seminar after seminar, in essay after essay.

Did Huxley truly think he was engaged on a religious crusade? He did back then, and many have continued to think in this way down to the present. Huxley was quite explicit that he was seeking a new religion to supplant the old, Christian religion. From a letter he wrote just before the *Origin*, seeing religion and science forever at war: "Few see it but I believe we are on the eve of a new Reformation and if I have a wish to live thirty years, it is that I may see the foot of Science on the necks of her Enemies . . . But the new religion will not be a worship of the intellect alone" (quoted by Desmond 1997, 253). Huxley's readers, friends and foes, saw what he was about, starting with the fact that he called his essays "Lay Sermons": "He has the moral earnestness, the volitional energy, the absolute conviction in his own opinions, the desire and determination to impress them upon all mankind, which are the essential characteristics of the Puritan character. His whole temper and spirit is essentially dogmatic of the Presbyterian or Independent type, and he might fairly be described as a Roundhead who had lost his faith." He shows all the signs: "the hortatory passages, the solemn personal experiences, the heart-searchings and personal appeals that are found in Puritan literature" (Baynes 1873, 502).

This making of Darwinian evolutionary theory into a kind of secular religion—"Darwinism"—continued down through the twentieth century (Ruse 2005). Thomas Henry's grandson, Julian, wrote a book called *Religion without Revelation*. Like his grandfather, he wanted to conquer conventional religion, not to eliminate it but to turn it to his own ends. God must go, but what remains is vital: "if, finally, there be no reason for ascribing personality or pure spirituality to this God, but every reason against it; then religion becomes a natural and vital part of human existence, not a thing apart; a false dualism is overthrown; and the pursuit of the religious life is seen to resemble the pursuit of a scientific truth or artistic expression, as the highest of human activities" (Huxley 1927, 53–54). More recently among today's evolutionists, Edward O. Wilson plays the same game of making a religion from his science. It is materialistic, presenting "the human mind with an alternative mythology that until now has always, point for point in zones of conflict, defeated traditional religion." Continuing: "Its narrative form is the epic: the evolution of the universe from the big bang of fifteen billion years ago through the origin of the elements and celestial bodies to the beginnings of life on earth. The evolutionary epic is mythology in the sense that the laws it adduces here and now are believed but can never be definitively proved to form a cause-and-effect continuum from physics to the social sciences, from this world to all other worlds in the visible universe, and backward through time to the beginning of the universe" (Wilson 1978, 192). We have a world vision, beyond the empirical, giving meaning to life. We have religion.

Rival Eschatologies

What of progress as the metaphysical backbone to Darwinism? Thomas Henry Huxley is an interesting case. He clearly did believe in progress in the social world. He worked all of his life for it. He believed also in progress in the biological world. At the same time, he saw that neither was straightforward. By the end of his life, the last decade of the nineteenth century, industry was stagnating, cities were cesspools of poverty and prostitution, rural England was staggering under the effect of imports from around the globe and more. In his personal life, Huxley had equal problems. A much beloved daughter died. Progress is possible, but it is struggle all of the way. We need to bring forward our cultural side, and use our intelligence and emotions to make for a happier, more peaceful tomorrow. The implication is that this can be done, but it will not be easy. Not entirely a sentiment shared by his grandson Julian, for all that the younger man inherited from the older man a lifetime's burden of depression. Julian Huxley was much more bullish about the chances of upward change. Acknowledging a debt to Spencer, Huxley wrote that evolutionary progress should be defined as "increased control over and independence of the environment" (Huxley 1942, 545). Thus defined, humans come top. Although, to quote Uncle Ben, with great power comes great responsibility. "The future of progressive evolution is the future of man. The future of man if it is to be a progress and not merely a standstill or a degeneration, must be guided by a deliberate purpose. And the human purpose can only be formulated in terms of the new attributes achieved by life in becoming human" (577).

Coming down to the present, Edward O. Wilson continues the tradition. "The overall average across the history of life has moved from the simple and few to the more complex and numerous. During the past billion years, animals as a whole evolved upward in body size, feeding and defensive techniques, brain and behavioral complexity, social organization, and precision of environmental control—in each case farther from the nonliving state than their simpler antecedents did" (Wilson 1992, 187). The enthusiasm is for absolute progress, leading to humankind. In his great book, *Sociobiology: The New Synthesis*, Wilson offers a tale of social evolution from colonial invertebrates, through the hymenoptera, onto the vertebrates and the primates, and now humans. "Man has intensified these vertebrate traits while adding unique qualities of his own. In so doing he has achieved an extraordinary degree of cooperation with little or no sacrifice of personal survival and reproduction." The big question is not whether progress occurred, but why it occurred. "Exactly how he alone has been able to cross to this fourth pinnacle, reversing the downward trend of social evolution in general, is the culminating mystery of all biology" (Wilson 1975, 382).

Two world pictures. Two religions. One, Christianity, with a creator God who cares for his favored creation, humankind. We unfortunately are tainted with sin and can be saved only through the free gift of God. Providence. The other, Darwinism, with a god who is nonexistent or indifferent about humans. We are, nevertheless, the pinnacle of evolution. We can moreover better our position through our own efforts. Progress. Rivals that we can fit into traditional apocalyptic categories. Traditionally, Christians have split on the future, the end of times (Ruse 2005). Revelation, the last book of the Bible, tells us that Jesus will come again, and it speaks also of a thousand-year period that will occur before the Day of Judgment. Some Christians, Saint Augustine notably, take all of this in at most a metaphorical fashion. For the saint, with the promises of Jesus, we are already in some sense living in the millennium. Don't look for a literal period in the future. Many Christians, today particularly evangelicals especially including Creationists, think that Jesus will come before the millennium, that there will be fighting (Armageddon), and then he will rule until Judgment Day. Other Christians, usually progressive thinkers like those in the Social Gospel movement at the beginning of the last century, think that we must work to create the millennium and only after this will Jesus reappear. With these latter two positions, particularly, we get very different doctrines about the proper course of behavior for Christians. Premillennialists tend to stress our sinful nature; the impossibility of doing things on our own; and hence the need to prepare for the Second Coming, through faith-based activities including attempts to convert others. Postmillennialists tend to stress the possibilities of change and of our using our God-given talents to improve the world and the place of humans within. Expectedly, postmillennialism as a general philosophy slops over into secular movements and thinking. For instance, the British Labor Party always concludes its annual general meeting by singing William Blake's hymn, "Jerusalem," published in 1808.

Bring me my bow of burning gold
Bring me my arrows of desire
Bring me my spear! Oh, clouds unfold!
Bring me my chariot of fire

I will not cease from mental fight
Nor shall my sword sleep in my hand
Til we have built Jerusalem
In England's green and pleasant land!

Without intending or implying that Darwinians are necessarily implicitly or explicitly Christian because they are so clearly postmillennialists, we can certainly allow that their religious position places them in this tradition.

Somewhat cynically, one might say that the real mark of religion is that you hate everyone who does not believe exactly as you do. Certainly, just as so many Christians distrust and dislike Darwinism, so many Darwinians return the compliment about Christianity. Thomas Henry Huxley was explicit in his views about the evolution-religion relationship. "Extinguished theologians lie about the cradle of every science as the strangled snakes beside that of Hercules; and history records that whenever science and orthodoxy have been fairly opposed, the latter has been forced to retire from the lists, bleeding and crushed, if not annihilated; scotched, if not slain" (Huxley 1860, 52). Grandson Julian sang the same song, seizing right on the (to him) offensive notion of Providence. Note, and we shall encounter similar sentiments later, the big objection is that the theology leads to complacency and to more ill than good. "Divine Providence is an excuse for the poor whom we have always with us; for the human improvidence which produces whole broods of children without reflection or care as to how they shall live; for not taking action when we are lazy; or, more rarely, for justifying the action we do take when we are energetic. From the point of view of the future destiny of man, the present is the time of clash between the idea of Providentialism and the idea of humanism—human control by human effort in accordance with human ideals" (Huxley 1927, 18).

Wilson is less hostile toward Christianity, but he too sees it as a rival. "I see no way to avoid the fundamental differences in our respective worldviews." On the one hand, Christianity. "You are a literalist interpreter of Christian Holy Scripture. You reject the conclusion of science that mankind evolved from lower forms. You believe that each person's soul is immortal, making this planet a way station to a second, eternal life. Salvation is assured those who are redeemed in Christ." On the other hand, Darwinism. "I am a secular humanist. I think existence is what we make of it as individuals. There is no guarantee of life after death, and heaven and hell are what we create for ourselves, on this planet. There is no other home. Humanity originated here by evolution from lower forms over millions of years." We are fancy apes who have adapted rather well to life here on

Earth. And this means that spiritual explanations and understandings are otiose. The same is true of behavior. "Ethics is the code of behavior we share on the basis of reason, law, honor, and an inborn sense of decency, even as some ascribe it to God's will" (Wilson 2006, 3–4).

Seminal Fluid

Let me hammer one final nail into this plank. In the context of Christianity and Darwinism, Providence and progress, premillennialism (especially among the haters of Darwinism) and postmillennialism, let us see how it all plays out in the different attitudes to a matter of great concern to the Victorians. A matter, as I can attest personally, of continuing concern to the headmasters of English boarding schools in the 1950s. I refer to that topic for hushed tones—shades of Dr Strangelove!—vital bodily fluid. Everyone knew that there were medical issues here. There is only so much fluid that a man can produce, and either it goes to making sex cells that flow from the loins or it goes to making bigger and better brains. "The seminal fluid is the very essence of the vital principle;—the most essential part of the blood; it is, in fact, the embryo of the species;—hence the frequent repetition of the vice above described, produces a wanton waste and overflow of this most nutritious secretion, and brings on all of the evils, which we shall further dwell on . . ." (Brodie 1845, 11). Not to keep the reader in suspense, these include "nervousness, hypochondriacism, depression of spirits, melancholy mania, epilepsy, hysteria, paralysis, dimness of sight, difficulty of hearing, etc."

Christians obsessed about all of this, and the dangers of waste—"the vice above described." Masturbation! Self-abuse! It wasn't just a medical issue; it went to the heart of their religion. It was the sin of Onan, except of course it wasn't, for his sin was refusing to impregnate the wives of his dead brother. But it was all about spilling seed on the ground, so one supposes that in the end it was the same thing. It was one of the lynchpins in the Christian case for original sin of the most filthy and disgusting kind. My headmaster stood in a long tradition of academic fearmongers. The Reverend Frederic W. Farrar was an enthusiast about the evolution of language, to the extent that Darwin proposed him for the Royal Society and, in turn, he was a pallbearer at Darwin's funeral. He was a master at Harrow School and then the sometime headmaster of Marlborough College—both English, private, boarding schools, misleadingly known as "public schools"—and later Dean of Canterbury. Farrar was also the author of

Eric or Little by Little, a boy's story that rivaled *Tom Brown's Schooldays* in popularity—and was incidentally the object of great scorn in Rudyard Kipling's schoolboy stories, *Stalky and Co.* Perhaps Stalky and chums were too far gone to appreciate its message, for it tells the dreadful story of a lad of great promise who sinks lower and lower, to be released finally only by death.

The problem was known to every reader. Eric was a wanker. "Kibroth-Hattaavah! Many and many a young Englishman has perished there! Many and many a happy English boy, the jewel of his mother's heart—brave, and beautiful, and strong—lies buried there. Very pale their shadows rise before us—the shadows of our young brothers who have sinned and suffered." The gaunt bodies of exhausted masturbators crowd the Empire. "May every schoolboy who reads this page be warned by the waving of their wasted hands, from that burning marle of passion where they found nothing but shame and ruin, polluted affections, and an early grave" (Farrar 1858, 86). Masturbation is not just a physical problem; it is a moral problem too. Kibroth-Hattaavah is the place where the Lord God made the Israelites "vomit through their nostrils" from meat eating, a foodstuff they had demanded getting, perhaps understandably, rather tired of manna for every meal. It denotes lust, and no reader would have thought that Farrar was referring here to quail, the source of the overabundance of meat that the Lord provided.

Turn now to the evolutionists. They illustrate perfectly the point I am trying to make, for they seized with no less enthusiasm on the limited supplies of available vital bodily fluid. They shared the Christian beliefs about the effects of misuse. Christians and Darwinians alike agreed completely that, in Charles Dickens's *David Copperfield*, the vile character of Uriah Heep—clammy hands, sunken chest, corrupted moral standards—was obsessing in unmentionable, ejaculatory ways over the spotless Agnes Wickfield. But whereas the Christians infused the topic with original sin, the Darwinians saw evidence supporting thoughts of social and cultural progress, reflecting biological progress. For them, the issue was not so much that of the solitary vice after lights out, but the rather the non-solitary vice after lights out. It is not masturbation that is the problem, but unrestrained heterosexual intercourse. Darwin in the *Descent* led the way, pushing on beyond the individual to the racial and social. The Irish have lots of kids and the Scots but few. Fortunately, nature takes a hand and it seems that the bigger-brained may have fewer children but overall do better at raising them and so things are tilted in the right direction. Apparently "the intemperate suffer from a high rate of mortality, and the extremely profligate leave few offspring" (Darwin 1871, 1, 174–175).

Masturbation or excessive copulation, it is all the same physiological process. Seminal fluid is a precious and limited commodity whether you be Christian or Darwinian. Yet, while for the Christians this all pointed to the end of Empire, for the Darwinians it pointed to the biological success of the big-brained, and the degeneration and extinction of the Epsilon-Minus Semi-Morons. "The poorest classes crowd into towns, and it has been proved by Dr. Stark from the statistics of ten years in Scotland, that at all ages the death-rate is higher in towns than in rural districts, 'and during the first five years of life the town death-rate is almost exactly double that of the rural districts.' " In addition: "With women, marriage at too early an age is highly injurious; for it has been found in France that, 'twice as many wives under twenty die in the year, as died out of the same number of the unmarried.' The mortality, also, of husbands under twenty is 'excessively high.' " And if this were not enough, "if the men who prudently delay marrying until they can bring up their families in comfort, were to select, as they often do, women in the prime of life, the rate of increase in the better class would be only slightly lessened" (Darwin 1871, 1, 175).

Same facts or "facts," seized on by both sides, but whereas one side makes them the foundation of pessimistic thoughts about original sin and personal degeneration, the other side makes them the foundation of optimistic thoughts about the arrival of superior human beings. World pictures infused with morality and social dictates and restrictions. Forecasts about what will happen to the good and to the bad. That is what religion is all about, and I rest my case about Christianity and Darwinism being rival religions. On that note, let us turn to our topic, the nature and causes of warfare.[1]

[1] What about my fellow philosophers who don't much care for Darwinism? This is a long-standing tradition in analytic philosophy, going back to the beginning of the twentieth century and the neo-Platonism of Bertrand Russell and G. E. Moore, and then a little later to Ludwig Wittgenstein, who was almost certainly reacting against some of the Germanic thinking we will encounter in later chapters (Cunningham 1996, Ruse 2017b). Specifically, in the context of this discussion, some like Alvin Plantinga (2011) are Christian, and it is clear that they share their fellow religionists' hostility toward Darwinism and for the same reasons. In the case of nonbelievers like Thomas Nagel (2012) and Jerry Fodor (with Piattelli-Palmarini 2010), it seems more a question of rival secular religions. For Nagel, a kind of neo-Aristotelianism—for some perhaps Marxism (Levins and Lewontin 1985). There is a feeling that Darwinism is too harsh and reductionistic—metaphors like "selfish gene" are thought offensive. A more group-friendly perspective is sought (Sober and Wilson 1998).

3 | Two Visions of War

Bent double, like old beggars under sacks,
Knock-kneed, coughing like hags, we cursed through sludge,
Till on the haunting flares we turned our backs,
And towards our distant rest began to trudge.
Men marched asleep. Many had lost their boots,
But limped on, blood-shod. All went lame; all blind;
Drunk with fatigue; deaf even to the hoots
Of gas-shells dropping softly behind.

Gas! GAS! Quick, boys!—An ecstasy of fumbling
Fitting the clumsy helmets just in time,
But someone still was yelling out and stumbling
And flound'ring like a man in fire or lime.—
Dim through the misty panes and thick green light,
As under a green sea, I saw him drowning.

In all my dreams before my helpless sight,
He plunges at me, guttering, choking, drowning.

If in some smothering dreams, you too could pace
Behind the wagon that we flung him in,
And watch the white eyes writhing in his face,
His hanging face, like a devil's sick of sin;
If you could hear, at every jolt, the blood
Come gargling from the froth-corrupted lungs,
Obscene as cancer, bitter as the cud
Of vile, incurable sores on innocent tongues,—
My friend, you would not tell with such high zest
To children ardent for some desperate glory,

The old Lie: Dulce et decorum est
Pro patria mori.

(Wilfred Owen, written October 1917)

I TURN NOW TO THE TASK in hand. War is a good case study for my claims about Darwinism as religion. Systematically, from the time of the *Origin* to the present, I shall look first at Christian thinking on war and then at Darwinian thinking on war, using the former as a template for the latter. If, through this comparative approach, I can show the kind of thinking of Darwinians (broadly construed) mimics the kind of thinking of Christians (broadly construed), moving beyond the strict science in directions properly called "religious," then I have made my case. Note that in focusing on what one might call the style or mode of thinking, I am far from ignoring or downplaying the huge differences between Christians and Darwinians over the nature, content, and meaning of war. Indeed, it is my overall case that the unease about Darwinism comes because it is arguing in the same domain as Christianity and yet to such different ends. Darwinism and Christianity can clash and they do clash. It is Niebuhr over and over again.

In the *Critique of Practical Reason* (1788), Immanuel Kant said: "Two things fill the mind with ever new and increasing admiration and awe, the oftener and more steadily we reflect on them: the starry heavens above me and the moral law within me." Epistemology, what we know and how. Ethics, what we should do and how. My inquiry is epistemological, but as I explained in the *Prolegomenon*, my motive is ethical. Since I was a schoolboy, I have been haunted by Wilfred Owen's terrible poem, written as the conflict was about to enter its final year. This book is a tribute to those who fought, those who suffered, those who died, those who were left behind.[1]

Christianity

War is conflict between societies or, derivatively, within societies. It is at some level sanctioned by the state—or by groups within the state—and it allows, insists, on behavior that does not happen in peacetime. Most notably,

[1] Just before the First World War, my father's mother (a teenager) died of tuberculosis. Abandoning their infant son, her husband, my father's father, my grandfather, went off to war, was badly hurt, and never returned to his family. My father was informally adopted by a childless couple and he learnt his true identity only when, as a teenager about to start work, he had to get an insurance card. One does not have to be unduly Freudian to suspect that my father's pacifism was a function of this rejection by his physically and emotionally damaged, professional-soldier father. One suspects also that stories of this kind, of families irretrievably fragmented and the subsequent suffering, can be repeated endlessly.

killing other human beings. It is common and widespread in, if not all, many or most societies. As I write, the United States is engaged in a drawn-out conflict in Afghanistan, with little prospect of an end. This is not a new state for the country. War began with the American Revolution, continued with war against Britain in Canada, then the horrific Civil War, more war with Spain at the end of the nineteenth century, and so to the twentieth—the First World War, the Second World War, Korea, Vietnam, and then two wars in Iraq. This is not even to mention the Cold War of the 1950s and later.

Not all religions are inherently antiwar. Islam, for instance, has always reflected its origins in a war-sanctioning culture (Kelsay 2007). From the first, however, war has posed special problems for Christians. The founder apparently unambiguously prohibits war and all related conflicts. Jesus said:

> Ye have heard that it hath been said, An eye for an eye, and a tooth for a tooth:
>
> But I say unto you, That ye resist not evil: but whosoever shall smite thee on thy right cheek, turn to him the other also. (Matthew 5:38–39)

That seems categorical. At least it did to the early Christians. "Love your enemies, bless them that curse you, do good to them that hate you, and pray for them which despitefully use you, and persecute you" (5:44). Violence is proscribed. The Lord's demand is that one refrains from war. One must be a pacifist. So, for instance, wrote the Alexandrian Church Father, Origen (ca. 185–ca. 254). "He nowhere teaches that it is right for his own disciples to offer violence to anyone, however wicked. For He did not deem it in keeping with such laws as His, which were derived from a divine source, to allow the killing of any individual whatever" (Origen, *Against Celsus*, in Holmes 2005, 48). None of this seems very practical, but the Christians were still a minority, unimportant sect, and they were thinking in apocalyptic terms. The Lord is returning soon and whatever happens down here is short term and will be corrected if need be.

Augustine

As the years went by, increasingly it became clear that the end is not in immediate sight and that Christians had better get used to living more in society rather than thinking that all things physical and social can be ignored in favor of the spiritual. Moreover, Christianity spread until finally under Constantine (306–337 AD) it moved toward being the state religion. This meant that Christianity and the state—a state that maintained an army that was willing to go to war—had to find some kind of modus vivendi. This they did, thanks above all to Saint Augustine (354–430), Bishop of Hippo in North Africa. Many of the early Christians had tackled the problem of war by downplaying the Jewish scriptures, the Old Testament, feeling that these writings were superseded by the teachings of Christianity, the New Testament. This meant that one could discard the bloodthirsty tales of the past, brought on or at least approved by the Jewish God. Jesus brought in a new moral realm, where there is no place for violence. Augustine wanted none of this rejection. For him, the Old Testament, Genesis particularly, was crucial, for it was through this that one could make sense of the death of Jesus on the Cross. Because of Adam and Eve, we are tainted by sin. Adam's fall transmits to us. "Wherefore, as by one man sin entered into the world, and death by sin; and so death passed upon all men, for that all have sinned" (Romans 5:12). The newborn child is pure only in the sense that he or she has not yet sinned, but the child has the tendency to sin and will surely do so before long. Jesus died as a sacrifice to make us pure (or at least declared pure) again. Substitutionary atonement! "For as by one man's disobedience [Adam] many were made sinners, so by the obedience of one [Jesus] shall many be made righteous" (Romans 5:19).

This central theological insight about ill-doing—"original sin"—has claim to being the most crucial notion in our whole discussion. So understand that the idea—note, from Paul not Jesus—refers not just to the act of wrongdoing, but also to its cause and explanation.[2] It is not part of our essential nature, our ontology as it were. It is something laid on us, making sin empirically bound to happen but not necessary in any logical or material sense—predictively certain but not inevitable in the sense of logically inescapable. Note the implications. Jesus's dying on the Cross could make no difference to our appetite. That is part of our animal nature. The death on the Cross can make a difference to our sinful nature. One should add parenthetically, that there is a difference between Catholics and Protestants on all of this (Adams 1999, 88). For Catholics, original sin is part of the loss of something above and beyond human nature, "original righteousness." Our nature was not affected by the action of Adam. For

---

[2] I shall often distinguish the Christianity of Jesus from that of Paul/Augustine. This is the thesis of Friedrich Nietzsche in *The Anti-Christ*. While I came to my conclusion independently, it would be churlish not to acknowledge my gradual change of mind about Nietzsche, from someone usually wrong and always unpleasant to the conviction that he was one of the most penetrating and original thinkers of the nineteenth century. This should not be taken as endorsing everything he says about Christianity—nor, I might add, endorsing everything he says about Darwin.

Protestants, it was rather that Adam's sin led to a corruption of our nature. Perhaps these differences are not as great as they seem, for the death on the Cross could speak only to the corruptions, and not to the original pre-sin human nature.

Ignore these niceties. Important is that the idea of original sin led naturally to Augustine's discussion of war. Far from rejecting the Old Testament, it became an important part of his position. War is to be expected. Why? Because we are sinners and war is a manifestation of our corrupted nature. "What is the evil in war? Is it the death of some who will soon die in any case, that others may live in peaceful subjection? This is mere cowardly dislike, not any religious feeling. The real evils in war are love of violence, revengeful cruelty, fierce and implacable enmity, wild resistance, and the lust for power, and such like" (Augustine, *Reply to Faustus the Manichean*, in Holmes 2005, 64). Because we are sinners and ongoing sinners, war will be with us always. God promised David peace and tranquility. "And I will appoint a place for my people Israel, and will plant him, and he shall dwell apart, and shall be troubled no more" (2 Samuel 7, 10). This is not for this world. "But if anyone hopes for so great a good in this world, and on this earth, his wisdom is but folly. Can anyone suppose that it was fulfilled in the peace of Solomon's reign?" (Augustine 413–426, 801).

Because of original sin, war is part of life. How then as Christians can we come to terms with this? In the purely religious (Christian) sense, Augustine thought that the Bible gives support for the role of the soldier and the necessity sometimes of being at war. War is an evil thing, but waging it can be a morally and religiously good thing. Augustine was not in the business of separating a private morality (against war) from a public morality or nonmoral realism (pro war). In not rejecting the Old Testament, Augustine felt free to make much of King David and his warlike mode of life. "Do not feel it is impossible for anyone to please God while engaged in active military service. Among such persons was the holy David, to whom God gave so great a testimony" (Augustine, Letter 61, in Holmes 2005, 61). Remember the Psalms: "Plead *my cause*, O LORD, with them that strive with me: fight against them that fight against me" (35:1). "Blessed be the LORD my strength which teacheth my hands to war, and my fingers to fight" (144:1). In the New Testament, there is the example of the centurion. He said, " 'I am a man under authority, and have soldiers under me: and I say to one, do this, and he doeth it,' Christ gave due praise to his faith; He did not tell him to leave the service" (Augustine, *Reply to Faustus*, in Holmes 2005, 64–65).

In treating more philosophically of war, Augustine drew heavily on the Roman thinker and orator Cicero (106–43 BC), who made central the notion of "natural law," meaning that which is right and proper is so because it is part of nature. "Nature has endowed every living creature with the instinct of self-preservation, of avoiding what seems likely to cause injury to life or limb, and of procuring and providing everything needful for life—food, shelter, and the like" (Cicero, *De Officiis*, in Holmes 2005, 25). It is not only that we do in practice do this, but it is right and proper that we do this. However, in the case of humans there is a big qualifier. We, uniquely, are the beings with reason. That gives us the ability to think not just about ourselves, but about others, our families, and our societies. Unfortunately, societies are going to quarrel and have different aims. As rational beings, we should always try to discuss and argue out differences without resorting to violence. War is the last measure, to be entered into reluctantly. Even then, it is constrained by natural law.

Augustine picked right up on this kind of thinking and, using our reasoning abilities as the way of converting Cicero's animal explanation of aggression into the Christian notion of original sin, it governed his treatment of war. We never enter into war because we want to and do so only reluctantly. This does mean that, although an evil, going to war can be justified. Where Augustine made his major contribution was in starting to spell out the conditions under which war is justified (*jus ad bellum*) and under which war must be fought (*jus in bello*). The conditions for a justified war must always be governed by the desired endpoint of peace, where this means harmony and order, and this leads first to the need of proper authority to wage war—no private conflicts; second to the need to be just—unrestrained territorial aggression is ruled out; and third to right intention—no thrusts for power covered by phony excuses. Intention, with the ultimate aim of peace, segues naturally into *jus in bello*, conduct in war. Augustine makes it very clear that there are limits on what one can do and how one does it. He also makes it clear that he thinks this is a Christian matter, although obviously he is with Cicero in thinking that these matters (and how to determine these matters) are also questions of natural law. God, as creator of the natural, stands behind His laws.

The influence of Augustine was and continues to be massive. Not that everything stands still. There has been nigh two thousand years of development and extension of his ideas. Most influential was the Italian Dominican, living nearly a thousand years after Augustine, Saint Thomas Aquinas (1225–1274). He is important not only for his theology and philosophy, but because his is the official position of the Catholic Church.

Aquinas wrote extensively about war in the Augustinian tradition, as did many subsequent others, including the great Protestant Reformers, Martin Luther and Jean Calvin. Both insist on the legitimacy of warfare, although expectedly as one who trained as a lawyer, there is much concern by Calvin about the rules and regulations governing just wars. If rulers and states have been given the power to rule and "to maintain the tranquility of their subjects, repress the seditious movements of the turbulent, assist those who are violently oppressed, and animadvert on crimes, can they use it more opportunely than in repressing the fury of him who disturbs both the ease of individuals and the common tranquility of all; who excites seditious tumult, and perpetrates acts of violent oppression and gross wrongs?" (Calvin 1536, IV, 20, 12). But doesn't this all go against biblical teaching? Calvin's response is that Jesus's message is not about any of this aspect of practical life, but more directed to our personal salvation. To naysayers: "I answer, first, that the reason for carrying on war, which anciently existed, still exists in the present day; and, secondly, that in the Apostolic writings we are not to look for a distinct exposition of those matters, their object being not to form a civil polity, but to establish the spiritual kingdom of Christ; lastly, that there also it is indicated, in passing, that our Saviour, by his advent, made no change in this respect." Augustine's authority is invoked for this last point.

The earlier Christian pacifist tradition never vanished. Particularly after the Reformation, with the return to the authority of the Bible—*solar scriptura*—there have been important groups who have turned from war. The Mennonite *Schleitheim Confession* of 1527 explicitly includes renunciation of the sword. Augustinian sin has not been binding. The Christianity has been that of Jesus of the Gospels, not Paul of Romans. At the end of the eighteenth century, in the spirit of the Enlightenment, the great Immanuel Kant—despite (or perhaps because of) his intense (and ever-influential) Lutheran Pietist background—repudiated the Augustinian notion of original sin. He did not let us off the hook. He thought we are "radically evil." The important thing is that it cannot be a "predisposition" to evil, which is something innate and brought on us—as in the Augustinian story—but is a "propensity" to evil, for which we ourselves as free beings are morally responsible. "Now this propensity must itself be considered as morally bad, so not as a natural predisposition but rather as something the man can be held accountable for; and consequently it must consist in unlawful maxims of the will" (Kant 1793). However, as we have a propensity to evil, so also we have the freedom to move to good. Note therefore that for Kant, and others in like traditions, freedom is absolute. We choose between right and wrong. Often we choose wrong, but entirely from our own doing and not because of original sin. To Augustinians, this is known (after a British monk, a contemporary of Augustine) as the Pelagian heresy.

Kant was not a pacifist. From his *Doctrine of Right*: "In the state of nature, *the right to make war* (i.e., to enter into hostilities) is the permitted means by which one state prosecutes its rights against another" (Kant 1785, 167). Continuing, "Thus, if a state believes it has been injured by another state, it is entitled to resort to violence, for it cannot in a state of nature gain satisfaction through *legal proceedings*." (See also Orend 2000, 48.) Famously, however, Kant wrote a spritely and influential essay arguing for the possibility of eternal peace. Anticipating what we shall see was the position of Herbert Spencer, Kant argued that trade and the like will make ongoing conflict seem less and less reasonable: "nature also unites nations which the concept of cosmopolitan right would not have protected from violence and war, and does so by means of their mutual self-interest. For the spirit of commerce sooner or later takes hold of every people, and it cannot exist side by side with war." A good end although don't be too pleased with yourself. "Thus states find themselves compelled to promote the noble cause of peace, though not exactly from motives of morality" (Kant 1795, 114). Others, particularly those who refuse to take up arms at all, while agreeing that peace here on earth is possible, tend to put things more in an eschatological context, arguing that, in the fullness of time, God will put things to rights. Today, the Jehovah's Witnesses refuse absolutely to have anything to do with war. During the Third Reich, this led to great suffering, even unto death. It was born with what one can truly say was Christian fortitude. One can tolerate a great deal of doorstep proselytizing because of this. Many, like my father, refuse to take up arms, but feel they can and should do something to aid the war effort. Quakers particularly, through the Friends' Ambulance Unit, have been leaders here. Ignoring internal differences, keep always in mind this alternative position. It will prove important.

Anglican Thinking

Move the discussion forward to Britain and the nineteenth century. The Church of England, the Anglican Church, is "established." This means it has a privileged position and role in English society and government. The monarch must be an Anglican, its leaders—the archbishops and senior bishops—are part of government, the House of Lords, and it was and still is the primary determiner of what is taught in the religious education classes in state-supported schools. Although it was her father, Henry the Eighth, who broke with Rome, the Church really came into being under Elizabeth, something made possible by the steadying influence of her very long reign (1558–1603). The Elizabethan Settlement was the compromise between the twin threats of continental Catholicism (and all of those annoying, evangelizing Jesuits determined on martyrdom) and the Calvinism of Geneva (where many English Protestants fled during the persecution under Elizabeth's older sister, Mary). It was then that the tradition of natural theology became firmly embedded in English Church practice and belief, a via media or middle way between the claims of authority of the Catholic Church and the *sola scriptura* of the Reformers. In style and buildings, Anglicanism is Catholic, as it is in the authority of the priest or parson, but in theology it is Calvinist. The doctrine of original sin is firmly embedded in the Thirty-Nine Articles—the statement of belief to which Anglicans subscribe. The Ninth Article begins: "Original sin standeth not in the following of Adam, (as the Pelagians do vainly talk;) but it is the fault and corruption of the Nature of every man, that naturally is engendered of the offspring of Adam; whereby man is very far gone from original righteousness, and is of his own nature inclined to evil, so that the flesh lusteth always contrary to the Spirit; and therefore in every person born into this world, it deserveth God's wrath and damnation." In the same spirit, we are told that we must obey the lawful civil magistrates. "*It is lawful for Christian men, at the commandment of the Magistrate, to wear weapons, and serve in the wars.*"

This all said, because it is the state church, and because Britain is a democracy not run by despots with the power to enact uniformity, Anglicanism has to be a wide tent able to include many varieties of belief, from the Anglo-Catholics near Rome in practice and belief to the Evangelicals (that probably influenced Darwin) with a strong emphasis on scriptural authority. In the mid-nineteenth century, the more central "broad church" was expectedly open to new theological and philosophical currents from the continent, particularly the so-called Higher Criticism, looking at the Bible as one would a secular document and demythologizing many of the more outlandish and miraculous claims, particularly those found in the Old Testament. This was combined with the traditional enthusiasm for natural theology, leading to support and promotion of science and technology. It is very much for this reason that, as we learned in chapter 1, despite some opposition, generally the Darwinian case for evolution was accepted without too much tension. Taking on new science was the English way. Of course, theological moves had also to be made. Two notorious collections of essays, *Essays and Reviews* in 1860 and *Lux Mundi* in 1889, captured much of the controversy and (to be fair) excitement. Showing that Bishop Wilberforce did not speak for a church united against evolution, in the first collection of essays, the Reverend Baden Powell, professor of geometry at Oxford University—and father of the scout master—was one of the first to speak out positively in favor of Darwin. Showing how far the church had moved by the end of the century, in the second collection of essays, the Reverend Aubrey Moore, Oxford fellow, argued that Christianity can not only live with Darwinism, it welcomes it. "Darwinism appeared, and, under the guise of a foe, did the work of a friend." Adding, "We must frankly return to the Christian view of direct Divine agency, the immanence of Divine power from end to end, the belief in a God in Whom not only we, but all things have their being, or we must banish him altogether" (Moore 1889, 268–269). Although these theologians were not progressionists as such, their thinking certainly inclined them a little this way. There was a tendency to echo the poet Alfred Tennyson who, a decade before the *Origin*, had been supposing that his dead friend Arthur Hallam, about whom he wrote *In Memoriam*, might have been of a higher type who turned up too soon.

> Whereof the man, that with me trod
> This planet, was a noble type
> Appearing ere times were ripe,
> That friend of mine who lives in God

Substitute Jesus Christ for Arthur Hallam, and you have a perfect fusion of Anglican theology and Darwinian science!

A state church by definition cannot isolate itself from the politics of the day. Around the time of Darwin, what about war? Between the battle of Waterloo in 1815 and the sailing of the British Expeditionary Force for France in the fall of 1914, there were only two significant wars—the Crimean (1854–1856) and the Boer War (1899–1902)—both in distant places and both limited in that they were fought only by professional soldiers. Expectedly, therefore, especially given the excitement over Darwin, war was not the chief focus of Anglican thought. However, although there were no other major wars, there was a bit more to the story than this. The Boer War, especially, was the tip of the iceberg, for particularly in the second half of the nineteenth century, Britain was engaged in many smaller wars, almost all associated with extension of or defense of the huge empire that the country was building. A third of the globe marked red because it fell within the British domain. Little surprise therefore that in our period of interest, from the mid-century on, when one does find religious writings and preachings on war, they reflect the immediate concerns of empire.

"Concerns of empire" were, as always—as still is as shown by the vote in favor of Brexit—embedded in the political philosophy of the British. Since the coming of William the Conqueror and the forging of the British state, defended and defined through such victorious conflicts as that against the Spanish Armada in 1588 and the French at Trafalgar in 1805, that island land has taken pride in and sustenance from its favored place in God's creation.[3]

> This royal throne of kings, this sceptred isle,
> This earth of majesty, this seat of Mars,
> This other Eden, demi-paradise,
> This fortress built by Nature for herself
> Against infection and the hand of war,
> This happy breed of men, this little world,
> This precious stone set in the silver sea,
> Which serves it in the office of a wall
> Or as a moat defensive to a house,
> Against the envy of less happier lands,—
> This blessed plot, this earth, this realm, this England.

By the mid-nineteenth century, Britain was not just a precious stone. Thanks to its empire, it ruled a good part of the land and sea elsewhere too.

Hence, even if it was far from the central focus of Anglican theological interest, something had to be said on the subject of war. Much appreciated were the thoughts of the mid-nineteenth-century professor of divinity at Oxford University, J. B. Mozley (among other things, brother-in-law of John Henry Newman). In an 1871 sermon on war, he invoked an

[3] When I was at school, the most vivid events of history were the great naval battles, up to and including the Second World War Battle of the Atlantic. We learned virtually nothing about the American War of Independence. Every year, the school would put on a Shakespeare play, almost always one of the histories. That way, you avoided the dangerous sexual stuff, such as the teenage shenanigans of Romeo and Juliet, and focused on the important stuff, like beating the hell out of the French at Agincourt. That said, we got excellent teaching on the last two centuries of European history. Otto von Bismarck was as familiar a figure as Benjamin Disraeli and William Gladstone and as sympathetically presented. If the second half of the century was to avoid the tragedies of the first half, we had to know what had gone so dreadfully wrong.

organismic vision of the state and used it as reason why a state, unlike an individual, is allowed to go to war. "The nation was one of those wholes to which the individual man belonged, and of which he was a part and member; it existed prior to Christianity, and was admitted into it with other natural elements in us; Christians were from the outset members of States; and the Church could no more ignore the State than it could the family" (Mozley 1871, 2). Continuing: "Individuals then are able to settle their disputes peaceably, because they are governed by the nation; but nations themselves are not governed by a power above them. This then is the original disadvantage under which nations are placed as regards the settlement of disputes; and in consequence of which, force takes the place of justice in that settlement" (6).

Harking back to Augustine. As Christians, there are times when we must take up arms. Fallen beings that we are, there can be no other option. Yet distinctively British. Almost in triumph, Mozley felt able to justify wars of colonial aggression and to give them a Christian backing. Apparently, there is a felt need for change. "There is doubtless an instinctive reaching in nations and masses of people after alteration and readjustment, which has justice in it, and which rises from real needs." So in a way, powerful nations from outside are doing everyone a favor. "Thus there is uneasiness in States, and an impulse rises up toward some new coalition; it is long an undergrowth of feeling, but at last it comes to the top, and takes steps for putting itself in force" (8). News to India and to much of Africa, I should imagine. Fortunately, salvation is at hand. "Strong States then, it is true, are ready enough to assume the office of reconstructors, and yet we must admit there is sometimes a natural justice in these movements, and that they are instances of a real self-correcting process which is part of the constitution of the world, and which is coeval in root with the political structure which it remedies." Adding: "And as Christianity at its commencement took up the national divisions of mankind, with war as a consequence contained in them, so it assumes this root of change and reconstruction with the same consequence as this fundamental tendency to re-settlement, this inherent corrective process in political nature" (9).

Then, as a complement as it were, moving over to the other of the ancient universities, we have the professor of philosophy at Cambridge, ordained member of the Church of England, Frederick Denison Maurice. Lecturing around 1870, he was nearly heady when it came to the fashioning and creative value of conflict. "We cannot forget that every Nation now existing in Europe became a Nation through war" (Maurice 1869, 203). Britain, by virtue of being part of the Roman Empire, not only lived well because of the connection, but was converted to Christianity before other pagan nations of Europe. Not that all went smoothly. There were relapses. "By battles—to what degree exterminating or subversive of the previous civilization historians may dispute—but certainly, by battles severe and bloody, the Saxons established their supremacy here. It seemed to the old inhabitants mere destruction, a relapse into barbarism and Paganism." Fear not, all will be made well. "We say that a mighty blessing came out of this apparent relapse." On the one hand, "a truer wholesome family life took the place of the corrupt family life which the Satirists of Rome describe and which passed from the capital into the provinces." On the other hand, "a people strong in the sense of neighbourhood, strong in the sense of personal existence, capable therefore of Law, of Government, bringing with them the roots of a vital native speech, overthrew colonists in whom there was a feeble sense of neighbourhood, a feeble feeling of personal responsibility" (203–204). All in all, the "Saxon wars, destructive as they might be, yet were in the strictest sense the commencement of a new life in our island." Providence, but—in the Anglican spirit of the times—an optimistic Providence one might say.

Darwin on Social Behavior

"A new life in our island." When Darwin was writing the *Descent*, we might expect to find—as I shall suggest we do find—that his thinking on war has debts to Anglican theology, starting with the obvious fact that any theology that goes back to Augustine is going to take natural law seriously. This begs for science. It also raises questions. Most particularly, we have seen that Darwin insisted that sociality (including morality) comes about through the evolutionary process of natural selection. Why then does one have to add war to the ingredient, especially given that when Darwin was writing this was all a quiet time on the war front? The answer is simple. War and peace, organized fighting and human sociality, are different sides of the same process, causally entwined in their genesis. It is hardly an exaggeration to say that perhaps the biggest debt to theology is the belief that in our post-lapsarian state, even if this be metaphorical, violence is going to be part of the human package. War is natural. War is at the heart of the evolutionary process. War led progressively to the appearance and nature of humans.

Expand on this thinking. First, Darwin stresses again that it is natural selection that has produced humans. "Had he not been subjected during primeval times to natural selection, assuredly he would never have attained to his present rank" (Darwin 1874, 142). Second, as with other animals, Darwin sees a very large number of factors as being involved. Several times, he mentions disease and illness as selective factors. Picking up on themes highlighted at the end of the last chapter: "In regard to the moral qualities, some elimination of the worst dispositions is always in progress even in the most civilised nations. . . . Profligate women bear few children, and profligate men rarely marry; both suffer from disease. In the breeding of domestic animals, the elimination of those individuals, though few in number, which are in any marked manner inferior, is by no means an unimportant element towards success" (137). War is clearly a major factor. "Savages, when hard pressed, encroach on each other's territories, and war is the result; but they are indeed almost always at war with their neighbours" (46). Expectedly from one who was now making much of sexual selection, women were often the immediate cause of all of this. "With savages, for instance the Australians, the women are the constant cause of war both between members of the same tribe and between distinct tribes. So no doubt it was in ancient times; 'nam fuit ante Helenam mulier teterrima belli causa' " (561).[4] Long before those Australians, as our intended Anglican parson would have known very well, there were the Ancient Jews.

> [10] When thou goest forth to war against thine enemies, and the LORD thy God hath delivered them into thine hands, and thou hast taken them captive,
>
> [11] And seest among the captives a beautiful woman, and hast a desire unto her, that thou wouldest have her to thy wife;
>
> [12] Then thou shalt bring her home to thine house, and she shall shave her head, and pare her nails;
>
> [13] And she shall put the raiment of her captivity from off her, and shall remain in thine house, and bewail her father and her mother a full month: and after that thou shalt go in unto her, and be her husband, and she shall be thy wife.
>
> —(DEUTERONOMY, 21:10–13)

Third, setting the discussion in the more general context, war does not preclude the evolution of social thoughts and behavior. "It is no

[4] This is a carefully bowdlerized version of the original Horace: "Nam fuit ante Helenam cunnus taeterrima belli causa." In English: "Even before the time of Helen, many a miserable cunny was the cause of war."

argument against savage man being a social animal, that the tribes inhabiting adjacent districts are almost always at war with each other; for the social instincts never extend to all the individuals of the same species" (108). Notice the other side to this equation. War does not negate the fact that we are social, but our sociality is of a kind that makes possible war. If we were born with an equal sense of moral inclination and obligation to every human on earth, then war might not be impossible, but it surely would be improbable. Although there are classic tales of sibling rivalry, starting with Cain and Abel, by and large one is less likely to take up arms against siblings than against strangers. David Hume knew the score: "A man naturally loves his children better than his nephews, his nephews better than his cousins, his cousins better than strangers, where everything else is equal. Hence arise our common measures of duty, in preferring the one to the other. Our sense of duty always follows the common and natural course of our passions" (Hume 1978, III, 2, i).

Now Darwin's discussion starts to get interesting, and very pertinent to our ends. Fourth, does Darwin see all of this evolution as a good thing, and is the consequence therefore that in some sense war is or was a good thing? Darwin always believed in social progress (Ruse 1996). Of course he did! He and his wife were the grandchildren of one of the most successful industrialists of the eighteenth century—the potter Josiah Wedgwood—and the rest of the family had not done too badly either. He was a successful man in Victorian Britain, the most successful society the world had ever known. That said, as we know, Darwin did not believe in any kind of necessary societal progress or, more pertinently, necessary biological progress. What he did believe in was some kind of biological equivalent to industrial struggle and consequent progress, anticipating what we saw in the first chapter are called biological "arms races." The implication simply has to be that in some sense, war was a good thing. It led to human beings. It may not be pleasant but then neither is going to the dentist. One of the quotes we have just had underlines the importance of war and like factors. "Had he not been subjected during primeval times to natural selection, assuredly he would never have attained to his present rank." And digging into details. "We might feel almost sure, from the analogy of the higher Quadrumana, that the law of battle had prevailed with man during the early stages of his development. As man gradually became erect, and continually used his hands and arms for fighting with sticks and stones, as well as for the other purposes of life, he would have used his jaws and teeth less and less" (Darwin 1874, 562).

This brings us to our fifth and final point, about where we are today. Darwin was no simple racist. He hated slavery. He was also a Victorian gentleman with the prejudices of his time and class. He was part of empire. Wedgwood crockery was going out to Canada and to Australia and to New Zealand and to South Africa and above all to India. His closest friend was the botanist Joseph Dalton Hooker, who in 1865 was promoted to succeed his father as director of the Royal Botanical Gardens at Kew, a post he held for twenty years. Kew was not some happy playground for kings and princes. It was a botanical factory, bringing in plants from all over the empire, raising them in greenhouses and then seeing if and how they could be sent back to other parts of the empire, to be grown profitably or to speak to immediate needs—breadfruit, coffee, tea, rubber, and more. It is little wonder then that Darwin accepted without question that Victorian upper-middle-class society was the apotheosis of civilization. The supreme good, to put things in Platonic terms.

Natural selection brought this about. To a (friendly) correspondent, almost at the end of his life, Darwin wrote: "I could show fight on natural selection having done and doing more for the progress of civilisation than you seem inclined to admit. Remember what risks the nations of Europe ran, not so many centuries ago of being overwhelmed by the Turks, and how ridiculous such an idea now is. The more civilised so-called Caucasian races have beaten the Turkish hollow in the struggle for existence. Looking to the world at no very distant date, what an endless number of the lower races will have been eliminated by the higher civilised races throughout the world" (Letter 13230, Darwin Correspondence Project, to William Graham, July 3, 1881). With the rise of civilization, the struggle falls away. "As man advances in civilisation, and small tribes are united into larger communities, the simplest reason would tell each individual that he ought to extend his social instincts and sympathies to all the members of the same nation, though personally unknown to him. This point being once reached, there is only an artificial barrier to prevent his sympathies extending to the men of all nations and races" (Darwin 1874, 122).

Civilized humans have transcended the struggle, but note how. "With civilised nations, as far as an advanced standard of morality, and an increased number of fairly good men are concerned, natural selection apparently effects but little; though the fundamental social instincts were originally thus gained" (137). We have transcended the struggle, which leads to the overall point—to give an argument we shall see repeated many times—today, the struggle as epitomized by warfare is positively counterproductive. On the one hand: "In every country in which a large standing army is kept up, the finest young men are taken by the conscription or are enlisted. They are thus exposed to early death during war, are often tempted into vice, and are prevented from marrying during the prime of life." Against this: "On the other hand the shorter and feebler men, with poor constitutions, are left at home, and consequently have a much better chance of marrying and propagating their kind" (Darwin 1874, 134).

Influences?

That is it. We evolved. We evolved through natural selection. War was part of that selective process. That doesn't mean we are not social. Sexual selection was involved. We came top, so in some sense war was a good thing. We civilized people have now transcended our biological past and war is no longer a good thing. Over the next one hundred and fifty years, there are going to be variations on this argument, but the pattern is set. Darwin is as important to Darwinism as Augustine is to Christianity. The *Descent of Man* is the secular response to the *City of God*. Although, as always with Darwin, one sometimes wonders "how secular"? Never forget that lurking Anglican background. By the time of the *Descent*, by the time of his discussion of war, Darwin's faith had faded to agnosticism. It remained an Anglican agnosticism, to coin an insightful oxymoron.

Insightful, as we see at once, when we turn to what in the *Descent* Darwin describes as "the ever memorable 'Essay on the Principle of Population,' by the Rev. T. Malthus." It was the sixth edition of 1826 on which Darwin had always relied. By this time, Malthus has introduced the escape clause of "prudential restraint," meaning that we can escape the struggle for existence through our own willpower. Savages don't and can't, but we more civilized people can and do. More than this, far from being the apotheosis of non-progressivist writing—a natural assumption given that the first edition of the essay was written explicitly against what Malthus through the false optimism of the Frenchman the Marquis de Condorcet and the Englishman William Godwin—at the end of the *Essay*, Malthus's natural theology takes over, and he argues that true improvement—God-given advance—is possible given the opportunities opened by such restraint. It's a tough world. It's God's world.

> However formidable these obstacles may have appeared in some parts of this work, it is hoped that the general result of the inquiry is such, as not to make us give up the improvement of human society in despair. The partial

> good which seems to be attainable is worthy of all our exertions; is sufficient to direct our efforts, and animate our prospects. And although we cannot expect that the virtue and happiness of mankind will keep pace with the brilliant career of physical discovery; yet, if we are not wanting to ourselves, we may confidently indulge the hope that, to no unimportant extent, they will be influenced by its progress and will partake in its success. (Malthus 1826, 543)

Struggle and then advance. There is more than a whiff of Anglian theology permeating the Darwinian discussion of war.

Understand, I am not now making silly claims that the *Descent of Man* on war is truly an exercise in either revealed or natural theology. Apart from anything else, there were other influences. The journalist Walter Bagehot was raised a Unitarian and, because he was unable to subscribe to the Thirty-Nine Articles, went to University College London rather than Oxford or Cambridge. In the years shortly before the *Descent*—in other words, after the *Origin*, the footsteps of which he was following explicitly—Bagehot published an influential series of articles, entirely secular, collected with the revealing title as *Physics and Politics: Or Thoughts on the Application of the Principles of "Natural Selection" and "Inheritance" to Political Society.* Showing again how misguided it is to talk of "non-Darwinian" revolutions, the theme is that societies evolve in a Darwinian fashion from militaristic groups to modern liberal institutions, from constraint and danger to freedom and opportunities to live a full and rewarding, non-threatened existence. "Whatever may be said against the principle of 'natural selection' in other departments, there is no doubt of its predominance in early human history. The strongest killed out the weakest, as they could" (Bagehot 1868).

The consequences are happily Darwinian also. "I think there is such a thing as 'verifiable progress,' if we may say so; that is, progress which ninety-nine hundredths or more of mankind will admit to be such, against which there is no established or organised opposition creed, and the objectors to which, essentially varying in opinion themselves, and believing one thing and another the reverse, may be safely and altogether rejected." We are invited to compare the English colonists down under to the indigenous Australian natives. No hesitation here about ordering. "Indisputably in one, and that a main sense, they are superior. They can beat the Australians in war when they like; they can take from them anything they like, and kill any of them they choose." It is not just fighting abilities that we are talking about here. We are talking also of the better things in life. "Indisputably in the English village there are more means of happiness, a greater accumulation of the instruments of enjoyment, than in the Australian tribe." It all comes about by moving from war to trade, where—and note this—war is now seen as a bad thing but formerly was a good thing. Comparing the age of primitive societies with today's age of advanced societies, "as the trade which we now think of as an incalculable good, is in that age a formidable evil and destructive calamity; so war and conquest, which we commonly and justly see to be now evils, are in that age often singular benefits and great advantages. It is only by the competition of customs that bad customs can be eliminated and good customs multiplied."

Expectedly, Darwin lapped this up. With the appearance of the articles on which Bagehot's book was based, he wrote recommending them to Hooker. "If you had time, you ought to read an article by W. Baghyot [Bagehot] in April no^{r} of Fortnightly, applying Natural selection to early or prehistoric politics & indeed to later politics—this, you know, is your view" (April 3, 1868, Darwin Correspondence Project, 6086). There are favorable references to Bagehot dotted through the *Descent*. As there are to the explorer and writer Winwood Reade—author of *The Martyrdom of Man*, a kind of secular, social Darwinian history of humankind—whose novel *The Outcast* made clear the overall Darwinian value of war. "At first, every step in the human progress was won by conflict, and every invention resulted from calamity. The most odious vices and crimes were at one time useful to humanity, while war, tyranny, and superstition assisted the development of man" (Reade 1875, 18). All of which leads to the proper conclusion that Anglican natural theology gave important aspects of the background for Darwin, including his discussion of war, but this was then fleshed out both by Darwin's own thinking and by his reading in the contemporary literature. As always, Darwin was a great revolutionary. He was not a rebel. In so many respects, he and his fellow Victorians thought the same way, sharing ideas back and forth, as he moved from Christian Providence to secular progress.

4 | Darwinism After Darwin

WHAT VARIATIONS ON THE DARWINIAN picture might we encounter? There are two fairly significant and serious ways of answering this question. Either way, we ask about the relationship between what we might call modern humans and ancient or protohumans—"hominins" in today's more technical language. You could, of course, argue that we simply have not changed much at all—bloody in the past, bloody in the present. We shall encounter this kind of thinking, but we can put it off until later. For now, focus on the more Darwinian picture of bloody in the past, peaceable in the present. The question now becomes whether this means that our biology has changed or at least it is something that comes out of our biology—in other words, we change biologically—or whether our biology stays the same and because of environmental factors, which may include our personal internal cultural environment, we become more peaceable—in other words, we change culturally. It is rather like quitting smoking. Did our genes change so we don't like cigarettes any more, or did societal pressures change so smoking is no longer accepted? It is all of a bit of a false dichotomy because you are not going to quit if your biology will not let you, but the main point is fairly obvious and meaningful. Without making too much of it, there is more than a whiff here of the Catholic-Protestant divide over original sin. Culture laid on biology (Catholic); biology itself changed (Protestant).

Herbert Spencer

Herbert Spencer (1820–1903) was very much one who stood on the biology side of the spectrum. He became an evolutionist around 1850 and since he never had a thought he didn't publish, at least three times, was soon churning out articles and books proclaiming the gospel. He was therefore over a decade behind Darwin in becoming an evolutionist (he was ten years younger!) but almost a decade ahead of Darwin in going public. He too was an enthusiastic, one might say overly enthusiastic, Malthusian. "We must call those spurious philanthropists, who, to prevent present misery, would entail greater misery upon future generations. All defenders of a Poor Law must, however, be classed among such" (Spencer 1851, 323). One is hardly surprised that one of Spencer's first published pieces postulated a form of natural selection, in the human case. First, the struggle: "Nature secures each step in advance by a succession of trials, which are perpetually repeated, and cannot fail to be repeated, until success is achieved. All mankind in turn subject themselves more or less to the discipline described; they either may or may not advance under it; but, in the nature of things, only those who do advance under it eventually survive." Second, the inference to selection: "it will be seen that premature death, under all its forms, and from all its causes, cannot fail to work in the same direction. For as those prematurely carried off must, in the average of cases, be those in whom the power of self-preservation is the least, it unavoidably follows, that those left behind to continue the race are those in whom the power of self-preservation is the greatest—are the select of their generation" (Spencer 1852, 499).

Note that Spencer so far anticipates the (public) Darwin that, although it was he who later was to provide "survival of the fittest" as an alternative term to "natural selection," he uses the word "select." Truly, however, a Darwinian-type solution to change was never Spencer's goal or commitment. He was quite prepared to use the struggle for existence and natural selection when it suited him, but at heart he was ever a Lamarckian—favoring the inheritance of acquired characteristics (Richards 1987). Like Darwin, Spencer started with the Malthusian struggle but whereas Darwin saw this as the key to selection, Spencer saw it as the key to effort and upward rise. Picking up on themes discussed earlier in this book, with the growth in brain size the fertility falls away and the Malthusian pressures are reduced. "Undue production of sperm-cells involves cerebral inactivity. The first result of a morbid excess in this direction is headache, which may be taken to indicate that the brain is out of repair; this is followed by stupidity; should the disorder continue, imbecility supervenes, ending occasionally in insanity" (Spencer 1852, 493).

This means that Spencer, like Darwin, saw a progressive rise up the chain of life, but unlike Darwin saw this progress as somehow prepackaged. His overall theory he referred to as the theory of "dynamic equilibrium," arguing that generally systems are in stasis and that, every now and then, such systems get disturbed and have to fight their way back to stasis, which tends to be of a higher form than previously. Showing ultimately the Germanic Romantic roots of his thinking, Spencer argued that progress involved a move from the simple, the homogenous, to the complex, the heterogeneous. Remember: "From the earliest traceable cosmical changes down to the latest results of civilization, we shall find that the transformation of the homogeneous into the heterogeneous, is that in which Progress essentially consists" (Spencer 1857, 2–3). You might think that this all implies that Spencer is going to be an enthusiast for warfare. Laissez faire at the intrastate level, ongoing conflict at the interstate level. This is simply wrong. The whole point about the move upward is that there is change, from massive reproduction to massive cognition. With this comes a falling away of violence and desire (or need) for strife.

Adding yet another idea to his eclectic stew, Spencer was an enthusiast for organicism, arguing that the state is an integrated organism, with parts serving the whole. This was very much part of his debt to German philosophy—which he got via the poet Samuel Coleridge—where there is heavy emphasis on the state as a unit. "[On the one hand] the state is absolutely rational inasmuch as it is the actuality of the substantial will which it possesses in the particular self-consciousness once that consciousness has been raised to consciousness of its universality. This substantial unity is an absolute unmoved end in itself, in which freedom comes into its supreme right. On the other hand, this final end has supreme right against the individual, whose supreme duty is to be a member of the state" (Hegel 1821, 258). Rather less metaphysically, Spencer allowed that there can be struggle within society, not to break it apart but to improve it. This does not mean that the state is exactly like the individual organism. The parts are physically separate in a way unlike biological organisms. "The parts of an animal form a concrete whole, but the parts of society form a whole which is discrete. While the living units composing the one are bound together in close contact, the living unit composing the other are free, are not in contact, and are more or less widely dispersed." This, however, is of no great concern, for language steps in to do the job: "though the members of a social organism, not forming a concrete whole, cannot maintain cooperation by means of physical influences directly propagated from part to part, yet they can and do maintain cooperation by another agency. Not in contact, they nevertheless affect one another through intervening spaces, both by emotional language and by the language, oral and written of the intellect" (Spencer 1860).

Spreading out from one society to the many—no doubt in part reflecting Quaker elements in his family history—Spencer was always against militancy, thinking that (real) arms races are a waste of time and money (Spencer 1904). Especially deplorable, at the end of the nineteenth century, was a major arms race between the British and German navies. Spencer was also in favor of free trade, thinking that among other things this would encourage relationships between nations. Morality emerges and our duties are to ensure that this happens by removing barriers and facilitating the process. "Ethics has for its subject-matter, that form which universal conduct assumes during the last stages of its evolution" (Spencer 1879, 21). Continuing: "And there has followed the corollary that conduct gains ethical sanction in proportion as the activities, becoming less and less militant and more and more industrial, are such as do not necessitate mutual injury or hindrance, but consist with, and are furthered by, co-operation and mutual aid."

All of this Spencer tied in with an analysis of the "militant society" as opposed to the "industrial society." No prizes are awarded for making out which societies today represent which type. "Such traits of the militant type in Germany as were before manifest, have, since the late war [the Franco-Prussian War of 1870–71], become still more manifest." The key point is that militant societies are more homogeneous whereas industrial societies are more heterogeneous, and while militant societies may have been needed in the early stages of evolution, this is no longer the case. In good Spencerian fashion, the need of a struggle for existence will fade away. "But now observe that the inter-social struggle for existence which has been indispensable in evolving societies, will not necessarily play in the future a part like that which it has played in the past. Recognizing our indebtedness to war for forming great communities and developing their structures, we may yet infer that the acquired powers, available for other activities, will lose their original activities." He agrees that "without these perpetual bloody strifes, civilized societies could not have arisen, and that an adapted form of human nature, fierce as well as intelligent, was a needful concomitant"—so in a major way he underlines the value of war in the past—but now with the arrival of modern societies, "the brutality of nature in their units which was necessitated by the process, ceasing to be necessary with the cessation of the process, will disappear" (Spencer 1882, 242).

In short, like Darwin we start with conflict but end with peace, but unlike Darwin we stay in the realm of biology all of the time. It is anything but a turning of one's back on or transcending of biology. "For when, the struggle for existence between societies by war having ceased, there remains only the industrial struggle for existence, the final survival and spread must be on the part of those societies which produce the largest number of the best individuals—individuals best adapted for life in the industrial state" (Spencer 1882, 699).

Thomas Henry Huxley

Thomas Henry Huxley was a long-time friend of Herbert Spencer. It is quite possible that, in the 1850s, Huxley became an evolutionist because of Spencer's hammering away at him during Sunday afternoon walks on Hampstead Heath. They fell out in the 1880s over their different takes on the evolutionary process and its consequences. Huxley always had a darker view of the world than did the ever-optimistic Spencer. Toward the end of his life, reflecting the general feeling that society was not improving as it should—his biologist protégé E. Ray Lankester was pushing the theme of degeneration and his sometime schoolteacher student H. G. Wells was writing novels, like the *Time Machine*, all about future human decline—Huxley turned to human evolution in a very gloomy sort of way. We have evolved, it is true, and perhaps even we have reached the headship, but to what end? Huxley was an enthusiast for the philosophy of David Hume. He even wrote a book about him (Huxley 1879). Noted already is Hume's famous claim that you cannot go deductively from matters of fact to matters of morality—from "is" statements to "ought" statements. In this spirit, the philosophers of the day were very critical of Spencer, arguing that he was going illicitly from the course of evolution—up to humans—to the presumption that this was a good thing (Sidgwick 1876). It was an egregious example of what G. E. Moore (1903) was to call the "naturalistic fallacy." In fact, I suspect that Spencer would have responded robustly and perhaps not without reason (in line with his Germanic Romantic influences) that he did not see the world, especially the living world, as dead and lifeless—Cartesian *res extensa*—and that he did think there was value accruing in the evolutionary process.

For Huxley, Hume said it all. Any value to be found in human life and action—certainly any valuing of peace over war—had to be done in the face of evolution through natural selection rather than because of it. It is very far from the case that those attributes that lead to success in the struggle are necessarily those that we think good. More often, it is the opposite. "Man, the animal, in fact, has worked his way to the headship of the sentient world, and has become the superb animal which he is, in virtue of his success in the struggle for existence" (Huxley 1893, 51). Continuing: "For his successful progress, throughout the savage state, man has been largely indebted to those qualities which he shares with the ape and the tiger; his exceptional physical organization; his cunning, his sociability, his curiosity, and his imitativeness; his ruthless and ferocious destructiveness when his anger is roused by opposition." Those days are past: "in proportion as men have passed from anarchy to social organization, and in proportion as civilization has grown in worth, these deeply ingrained serviceable qualities have become defects. After the manner of successful persons, civilized man would gladly kick down the ladder by which he has climbed. He would be only too pleased to see 'the ape and tiger die.'" For Huxley, if war is to be overcome, and I am not at all sure that by the end of his life he thought that this was at all possible, then it must be by rising above biology and not by being carried along by it.

Parenthetically, it hardly needs pointing out that Darwin was not the only one drawing on Christian heritages. One might think that the father of agnosticism would be drawing on secular sources, like the writings of Thomas Hobbes (Sussman 2013, 100). Possibly so, but there were more significant influences. We have seen how his contemporaries saw the Puritan streak in Huxley's worldview and he himself wrote (to that friend to whom he was to declare his religious intent) of his own "Scientific Calvinism." With Darwin, for all that the older naturalist shared the belief that there is something innately violent in our nature, one senses a comfortable Anglican attitude toward sin, one that colors his quite optimistic view of human nature. Huxley is otherwise, and there is more than a hint of a heavy-duty, Augustinian original sin in his view of human nature and of facile beliefs about progress. "I doubt, or at least I have no confidence in, the doctrine of ultimate happiness, and I am more inclined to look the opposite possibility fully in the face, and if that also be inevitable, make up my mind to bear it also" (Letter, October 10, 1854, to Frederick Dyster). One has to take with a slight pinch of salt claims like these from a man who spent his whole life laboring to improve the lot of his fellows. One nevertheless senses the mood. Later in life, experience made Huxley even more blunt. To a correspondent trying to persuade him of an evangelical worldview, he wrote:

> As a matter of fact, men sin, and the consequences of their sins affect endless generations of their progeny. Men are tempted, men are punished for the

sins of others without merit or demerit of their own; and they are tormented for their evil deeds as long as their consciousness lasts.

The theological doctrines to which you refer, therefore, are simply extensions of generalisations as well based as any in physical science. Very likely they are illegitimate extensions of these generalisations, but that does not make them wrong in principle.

—(LETTER TO THE REVEREND C. VOYSEY, November 18, 1876; in Darwin 1985–, 24, 345)

While Huxley may have been a little extreme, it is worth keeping in mind the extent to which Augustinian views on sin lurk beneath many Darwinian treatments of human nature and of how it leads to war. It is also worth keeping in mind the extent to which it was Huxley, above all others—certainly above Darwin—who was going out night after night until almost the end of the century, preaching this sort of stuff to social groups and clubs, to the general public, as well as writing one frenetic essay after another for the popular or semi-popular press. Talk about savages and tigers and apes was just the stuff for the troops, to use a phrase.

Alfred Russel Wallace

Look at a couple of outliers on this problem—outliers in their science, not in the shared horror at war and the determination that it can and must be ended. Alfred Russel Wallace (1823–1913) was several years younger than Darwin: although middle class, about as low in that category as Darwin was high. Wallace—who always seemed dogged by misfortune, starting with the fact that they could not spell his middle name correctly on his birth certificate—used to joke that his solicitor father had no worries because there was no lower that he could sink (Wallace 1905). Employed as a surveyor early in life—there was much work because the railways were being laid everywhere—and being thus very conscious of the poverty of the rural poor, at the age of fourteen Wallace was taken to hear the Scottish mill owner and socialist Robert Owen. A lifelong convert, Wallace always looked upon selection as a group phenomenon—which could include unrelated members—as is hinted by the title of the essay he first sent to Darwin in 1858: "On the Tendency of Varieties to Depart Indefinitely from the Original Type" (Wallace 1858; see also Ruse 1980). For Wallace, therefore, war within a species was always an unnatural phenomenon, which led not at all to evolutionary change and certainly not to progress of any kind.

Like Darwin—indeed, inspiring Darwin—in major respects Wallace saw humankind as "escaping" natural selection. With this comes an increase in the hatred of conflict. "By his superior sympathetic and moral feelings, he [humankind] becomes fitted for the social state; he ceases to plunder the weak and helpless of his tribe; he shares the game which he has caught with less active or less fortunate hunters, or exchanges it for weapons which even the sick or the deformed can fashion; he saves the sick and wounded from death; and thus the power which leads to the rigid destruction of all animals who cannot in every respect help themselves, is prevented from acting on him" (Wallace 1864, clxviii). Why then war? Wallace saw war as a cultural or societal issue, but unlike Darwin it was not something from which we escape thanks to civilization, but something imposed upon us by civilization. "Every addition of territory, every fresh conquest even of barbarous nations or of savages, provides outlets and additional places of power and profit for the ever-increasing numbers of the ruling classes, while it also provides employment and advancement for an increased military class in first subduing and then coercing the subject populations, and in preparing for the inevitable frontier disputes and the resulting further extensions of territory." Wallace stressed that it is the nobs who get all of the benefits, and the proles get stuck with the bills: "they invariably suffer from increased taxation, either temporarily or permanently, due to increased armaments, which the protection of the enlarged territory requires" (Wallace 1899, 213).

In Wallace's opinion, there is in addition a nasty undercurrent of jingoism about the whole business. "We boast of our Empire on which 'the sun never sets'; and lose no opportunity of expressing our determination and vaunting our ability to keep it." Wallace would have none of this. "Again and again we have waged unjust wars, as those against China, Burma, Egypt, and North-West India; while our last exploit—the most unjust and disgraceful of all—was the conquest of the two Boer Republics, after a petty quarrel deliberately founded on fraud and aggression" (213). Wallace was no lover of British pomp and glory. We should be ashamed of our behavior in South Africa. Wallace was a strange man, much given to maverick enthusiasms, going strongly against the tide. To his great credit, he was fearless in his often-isolated opposition. That he was bucking the Anglican theology of Oxford and Cambridge would be a matter of pride, not despair. One might add parenthetically that, although Wallace was raised an Anglican, there seem to have been strong Quaker influences in his childhood (Shermer 2002, 48). Friends have never bought into Augustinian views on sin. We are sinners, but not in the innate way supposed by Catholics and the Lutherans and Calvinists. War is not natural. We can therefore escape it, if we take the right track.

Petr Kropotkin

Prince Petr Kropotkin was Russian, an aristocrat, the child of a father who owned three hundred serfs. He broke with his background—notoriously, today he is still the most famous anarchist ever—and ended up being arrested and sent to the Peter and Paul prison in St. Petersburg. Escaping, he ended in exile in England. Always interested in science, it was there that he formulated and published his theory of "mutual aid," something that reflected strongly his Russian heritage (Todes 1989). For people like Darwin and Spencer—Wallace also—the Malthusian population pressure was the starting point. They had a point. The way that numbers were rising was positively scary. The population of England doubled between 1781 (about seven million) and 1831 (about fourteen million). London alone grew from a million and a quarter in 1801 to over two million in 1831. The struggle for existence was known only too well.

In Russia, however, things were different. Whatever the exact causes, population growth was linked to the move to an industrial society. If nothing else, in rural areas with limited land resources, people tended to marry later and put off large families, whereas in urban areas, often it was children who could best work in factories—tending weaving machines and the like—and hence the pressure was on many kids, quickly. Russia was late to industrialize, and in any case the vast lands meant it was not space that was at a premium. Far more significant was the awful climate. Those harsh Russian winters. Where the real struggle occurred was between organisms and the environment and, to succeed, the feeling (and this was general among Russian biologists and not exclusive to Kropotkin) was that organisms had to hang together to have any chance of success. "In the animal world we have seen that the vast majority of species live in societies, and that they find in association the best arms for the struggle for life: understood, of course, in its wide Darwinian sense—not as a struggle for the sheer means of existence, but as a struggle against all natural conditions unfavourable to the species." These truly are natural and not human-made conditions, and hanging together is the only solution. "The animal species, in which individual struggle has been reduced to its narrowest limits, and the practice of mutual aid has attained the greatest development, are invariably the most numerous, the most prosperous, and the most open to further progress" (Kropotkin 1902, 158).

Hence, as with Wallace, although for somewhat different reasons, war was a disease of civilization, rather than something that civilization made otiose. "One single war—we all know—may be productive of more evil, immediate and subsequent, than hundreds of years of the unchecked action of the mutual-aid principle may be productive of good" (159). Kropotkin shared with the more central figures a hatred of war and a belief that it can be overcome, but differed from them in thinking that it is far from obvious that the simple growth of civilization gives all of the answers. With Wallace he believed to the contrary that war comes from unrestrained civilization. It is only by recognizing this and taking remedial steps that we can hope for a better future. Again, parenthetically, one might add that perhaps Wallace and Kropotkin were not so far apart in influences. Orthodox Christianity has never been Augustinian or bought into his overall theological picture of human nature.

America

From its founding, America has been a country of contrasts, and nowhere do we see this more clearly than when it comes to evolution and Darwinism. On the one hand, one of Darwin's closest scientific friends was Asa Gray, professor of botany at Harvard University. He was let into the secret about natural selection before the *Origin* was published, and then spent the next twenty years defending it (Gray 1876). Although there was scientific opposition to evolution—mentioned already was Gray's Harvard colleague, the Swiss-born, German-educated ichthyologist, Louis Agassiz (1859)—it was not long lasting. Through the latter part of the century, the exciting fossil finds from the American West fleshed out the story of evolution in a way undreamed of by either supporters or critics (Bowler 1984). In the twentieth century, American evolutionary biology went from strength to strength, with major ideas and empirical discoveries coming from the New World. Yet, on the other hand, there was the idiosyncratic evangelical Protestantism of the country, particularly in rural America and above all in the South, a religion that was firmly—obsessively—fixed on a literalistic reading of the bible, especially Genesis, and that hence rejected evolution in all of its forms, but above all in its Darwinian forms.[1] This climaxed in

[1] Opposition to evolution could also be found in poorly educated, lower or lower-middle-class, English-only-speaking, urban white people, threatened as they felt by the influx of Catholics and

1925 in the notorious Scopes Monkey Trial, when in Dayton, Tennessee, a young schoolteacher was put on trial for teaching his students that humans are descended from other primates (Larson 1997). Creationism, as it is generally known, was there before—even in the 1870s professors were losing their jobs for teaching evolution—and it was thereafter, even to the present (Numbers 2006). There have been several relatively recent court cases, notably in Little Rock, Arkansas, in 1981 and in Dover, Pennsylvania, in 2005, where Creationist forays into public education had to be fought back (Pennock and Ruse 2008). Today, thanks to the charter school movement, there is probably more state-supported teaching of Creationism than at any point since the Pilgrims arrived on America's shores.

Later we shall make brief mention of the Creationists, but for now we can leave them, for although they have been very influential in such things as molding the content of state-supported-school biology textbooks, they are not really representative of mainstream religion. Augustine was not a Creationist in the modern sense—as a sophisticated Roman, he thought it would be foolish to take literally the metaphors and images in which God spoke to the illiterate and science-ignorant Israelites. Hence, there was the need of interpretation of many of the more fantastical claims of the Old Testament. In this, Augustine has been joined by both Catholics and Protestants. Turning then to evolution, we find—understandably in a country that had just gone through a horrendous Civil War—issues of war and peace were of interest to American evolutionists, especially those who really took Darwin seriously. One of the most interesting—who can stand here as representative—is the Pragmatist philosopher and psychologist William James. He was an immediate convert to evolution and not just to evolution but to natural selection, and not just to natural selection but natural selection in the human world (Ruse 2009). Expectedly, therefore, we find James holding forth on the violent nature of our origins, and as expectedly in some sense it is implied that this was a good thing, at least inasmuch as it led to us. "*Pugnacity; anger; resentment.* In many respects man is the most ruthlessly ferocious of beasts." From the individual's viewpoint, this has good consequences. "Constrained to be a member of a tribe, he still has a right to decide, as far as in him lies, of which other members the tribe shall consist. Killing off a few obnoxious ones may often better the chances of those that remain. And killing off a neighboring tribe from

---

Jews and other foreigners from Europe and elsewhere. Origins and religions of incomers suitably adjusted, this is still true.

whom no good thing comes, but only competition, may materially better the lot of the whole tribe" (James 1890, 717).

You might think that this is the end of the story, because famously—perhaps, better, notoriously—Pragmatism has a reputation as a philosophy that endorses violence and conflict. Bertrand Russell (1948) spoke of it as a "power philosophy" since truth seems to be a matter of who wins, who is in charge. In the Great War, many thought it was with almost indecent haste that John Dewey, the great twentieth-century Pragmatist, endorsed Woodrow Wilson's declaration of war against Germany. In the words of one critic: "War has a narcotic effect on the pragmatic mind." Continuing: "The war has revealed a younger intelligentsia, trained up in the pragmatic dispensation, immensely ready for the executive ordering of events, pitifully unprepared for the intellectual interpretation or the idealistic focusing of ends. . . . There seems to have been a peculiar congeniality between the war and these men. It is as if the war and they had been waiting for each other" (Bourne 1917, 59–60). James, however, was a more sensitive thinker. Since he died in 1910, we cannot say how he would have behaved and argued in the upcoming war. It is far from certain that he would have jumped on the pro-war bandwagon with the vigor of Dewey. He accepted that we have inherited a pugnacious nature and at one level accepted that biologically speaking we are stuck with it. "Our permanent enemy is the noted bellicosity of human nature. Man, biologically considered, and whatever else he may be in the bargain, is simply the most formidable of all beasts of prey, and, indeed, the only one that preys systematically on its own species. We are once for all adapted to the military status" (James 1904, 121). James sounds almost like an Augustinian on the subject of original sin. "A millennium of peace would not breed the fighting disposition out of our bone and marrow, and a function so ingrained and vital will never consent to die without resistance, and will always find impassioned apologists and idealizers." One suspects one hears here echoes of the thoughts of someone coming to maturity during the Civil War.

Yet, James didn't think this was all that there is to be said, for he believed it might be possible to channel this aggression into more peaceful and altruistic activities. One hears echoes here of his father's Swedenborgian enthusiasms, for the theology of that movement stresses the individual's involvement and personal effort in the move from sin (Swedenborg 1771, 647). Certainly, James fits the pattern we have seen in the British evolutionists, thinking war led to our species and in some sense therefore must be considered good, but that it is no longer good. The question he asked, innovatively, is, given the legacy of the past, where do we go from here? This was the theme of his famous late-in-life essay: "The moral equivalent of war." James was unambiguously against war today, and he declared himself a pacifist. He did not share Christian convictions that war is bound to happen, now and forever more. "I devoutly believe in the reign of peace and in the gradual advent of some sort of socialistic equilibrium. The fatalistic view of the war function is to me nonsense, for I know that war-making is due to definite motives and subject to prudential checks and reasonable criticisms, just like any other form of enterprise" (James 1910, 170). Probably thinking of German militarism, he continued: "when whole nations are the armies, and the science of destruction vies in intellectual refinement with the science of production, I see that war becomes absurd and impossible from its own monstrosity."

One very interesting and innovative aspect of James's thinking was that he wanted to see war in a positive sense, then and now, not in the killing but in the qualities needed in the warrior, and how they might be both cherished and turned to more peaceful ends. "We must make new energies and hardihoods continue the manliness to which the military mind so faithfully clings. Martial virtues must be the enduring cement; intrepidity, contempt of softness, surrender of private interest, obedience to command, must still remain the rock upon which states are built" (James 1910, 170). This, famously, led James to suggest a kind of proto Peace Corps, where young people can channel the military virtues for the good of their society specifically and humankind generally. "If now—and this is my idea—there were, instead of military conscription, a conscription of the whole youthful population to form for a certain number of years a part of the army enlisted against Nature, the injustice would tend to be evened out, and numerous other goods to the commonwealth would follow. The military ideals of hardihood and discipline would be wrought into the growing fiber of the people; no one would remain blind as the luxurious classes now are blind, to man's relations to the globe he lives on, and to the permanently sour and hard foundations of his higher life" (171). For some reason, we academics seem always happy, eager even, to consign our young people to the hardest and most degrading of manual occupations. No white collar jobs for them! "To coal and iron mines, to freight trains, to fishing fleets in December, to dishwashing, clotheswashing, and windowwashing, to road-building and tunnel-making, to foundries and stoke-holes, and to the frames of skyscrapers, would our gilded youths be drafted off, according to their choice, to get the childishness knocked out of them, and to come back into society with healthier sympathies and soberer ideas" (171–172). There is something almost biblical in the enthusiasm—Abraham sacrificing Isaac to show God that he, Abraham, is a loyal servant. Likewise, the young in James's worldview. "They would have paid their blood-tax, done their own part in the immemorial human warfare against nature; they would tread the earth more proudly, the women would value them more highly, they would be better fathers and teachers of the following generation" (172). One assumes and hopes that, today, James would be open to the idea that atoning, civilian boot camp for the gander would also be atoning, civilian boot camp for the goose; and that young women might join such a corps and play a full and equal role. Why should misery be the exclusive privilege of the young of one sex alone?

James, as always, looks forward as well as backward. Ignore my teasing. Like Huxley, he is a truly great man. Perhaps, had he lived longer, Pragmatism would not be burdened with the stain of unbridled enthusiasm for conflict. Judged from our perspective, the important point is that he is with his fellow Darwinians in seeing war as having produced humankind, but as neither necessary nor inevitable in the years to come. More precisely, James is with his fellow Anglophone Darwinians in seeing war as neither necessary nor inevitable. Such was not the case on the continent.

German Philosophy

Cross the English Channel and move east to Germany. We are no longer dwelling in either the well-established Britain or the increasingly confident United States of America. Here, we find a very different climate. Germany at the end of the nineteenth century and into the twentieth was a country newly united and searching for an overall theory to justify and bind, at the same time blessed or rather cursed with an over-dominant military and a weak civilian government (Clark 2006). It is within this political and cultural milieu that we must find German philosophical and theological thinking about the issue of war. Germany did not have to be distinctive. Already we have encountered Immanuel Kant with his hopes of perpetual peace. Although it is self-interest that offers the hope of perpetual peace, Kant would not be Kant unless he saw the possibility of some moral gain. In a peaceful society, one might hope for moral improvement of its citizens: "we cannot expect their moral attitudes to produce a good political constitution; on the contrary, it is only through the latter that the people can be expected to attain a good level of moral culture. Thus that mechanism of nature by which selfish inclinations are naturally opposed to one another in their external relations can be used by reason to facilitate the attainment of its own end, the reign of established right" (Kant 1795, 113).

Kant's philosophical successor, professor in the heart of Prussia at Berlin, Georg Friedrich Wilhelm Hegel, would have none of this perpetual peace nonsense. Not only was it not achievable, it was not desirable. War, in some sense, is not so much a good thing or a necessary thing, but a morally healthy thing. "I have remarked elsewhere, 'the ethical health of peoples is preserved in their indifference to the stabilisation of finite institutions; just as the blowing of the winds preserves the sea from the foulness which would be the result of a prolonged calm, so also corruption in nations would be the product of prolonged, let alone "perpetual" peace'" (Hegel 1821, 324).

It is important to note that this is not an anticipation of the kinds of convolutions that the Darwinians were tying themselves into over whether or not war is a good thing. At most, Darwin would have said that, if we are to have humans, who are indeed a good thing, then war is a necessary thing. In a sense, it is a reluctant necessity, because, as soon as one can, the need is to get away from war. The idea that it was ethically healthy would have been alien and repugnant to Darwinians. Better we go down the coal mines or wash dishes or clothes! For Hegel and successors—as, to be fair, we have seen hinted in some of our Anglican theologians, and as we shall see stressed repeatedly in the war to come—not only was war a good thing, it was a good thing for us morally.[2] In context, particularly as we are building the best kind of state—as manifested by the growth and coming together of the parts to make the new Prussian-infused Germany. Thus the philosopher Fichte. "The noble-minded man's belief in the eternal continuance of his influence even on this earth is thus founded on the hope of the eternal continuance of the people from which he has developed . . . . Hence, the noble-minded man will be active and effective, and will sacrifice himself for his people" (Fichte 1808, 115). Again, the value of the group over the individual is prominent. "Life merely as such, the mere continuance of changing existence, has in any case never had any value for him; he has wished for it only as the source of what is permanent. But this permanence is promised to him only by the continuous and independent existence of his nation. In order to save his nation, he must be ready even

[2] Note that war is good only as a side-effect of something that is, in itself, essentially bad. If there were no war, then presumably there had been no original sin, and so no need of the healthy, bracing effects of war. The difference with Darwinism is that, for the latter, war is not essentially bad, even though it is bad now.

to die that it may live, and that he may live in it the only life for which he has ever wished" (136). Remember how for Darwin it is the individual first over the group, although he does then go on to show how individual interests can then go on to play out into group interests.

Friedrich von Bernhardi

The legacy of the holistic thinking of Hegel and others was bolstered by other forces, particularly the Volkish movement turning to a glorious (if mythical) medieval past, as represented by the fairy tales collected by the Grimm brothers and above all by the operas of Wagner, with their doings of mortals and immortals in a world now gone. We have a vision of the state as an organic unity, manifested above all in the German state (Harrington 1996). With this kind of mindset, it is no very great surprise that, in the years after the *Origin*, German thinkers, especially those associated with the all-powerful military, cherry-picked convenient aspects of Darwin's thinking and, ignoring that which was less convenient, used them to bolster and legitimize their case. Above all, coming just before the outbreak of the Great War, there was the German general, a member of the Grand General Staff, Friedrich von Bernhardi. His *Germany and the Next War* laid things out starkly. "War is a biological necessity of the first importance, a regulative element in the life of mankind which cannot be dispensed with, since without it an unhealthy development will follow, which excludes every advancement of the race, and therefore all real civilization. 'War is the father of all things' " (von Bernhardi 1912, 18). Happy to use the authority of Darwin—"The struggle for existence is, in the life of Nature, the basis of all healthy development" (18)—von Bernhardi knew full well that he was drawing on older and deeper sources. " 'To supplant or to be supplanted is the essence of life,' says Goethe, and the strong life gains the upper hand. The law of the stronger holds good everywhere." This does not mean that notions of progress are absent. Anything but. Indeed, that is the whole point. Evolutionists were picking up on notions of progress, and Spencer in particular was picking up on Germanic Enlightenment notions of progress, seeing upward change as inevitable in a way alien to Darwin's industrialized notion. Von Bernhardi was more Spencerian than Darwinian as he reached back comfortably to the beginning of the nineteenth century. "In the extrasocial struggle, in war, that nation will conquer which can throw into the scale the greatest physical, mental, moral, material, and political power, and is therefore the best able to defend itself."

Note how this is bound up with progress. "War will furnish such a nation with favourable vital conditions, enlarged possibilities of expansion and widened influence, and thus promote the progress of mankind; for it is clear that those intellectual and moral factors which insure superiority in war are also those which render possible a general progressive development" (20). He is Hegelian in his thinking about progress and the need for conflict. "Without war, inferior or decaying races would easily choke the growth of healthy budding elements, and a universal decadence would follow" (20). Progress, yes, but not really the progress of Darwin. Nor, to be fair, really that of Herbert Spencer who—as his thinking matured in the second half of the nineteenth century—in his urge for internationalism and peace truly sounds much like the object of Hegel's scorn, Immanuel Kant.

Underlining his uneasy relationship with Darwin, von Bernhardi knew the real opponent of Germany—Britain! Magnanimously in a book published at the beginning of the war—*Our Future: A Word of Warning to the German Nation*—which the translator transformed into—*Britain as Germany's Vassal*—he gave the conditions needed to avoid war between the two countries. For a start: "England would have to give Germany an absolutely free hand in all questions touching European politics, and agree beforehand to any increase of Germany's power on the Continent of Europe which may ensue from the formation of a Central European Union of Powers, or from a German war with France" (von Bernhardi 1914, 152). Since this would fly in the face of five hundred years of British diplomacy, von Bernhardi obviously knew that his suggestions would not be taken seriously. Nor would other conditions, like allowing Germany to do what it liked in Africa and Asia and not to hinder the rise and activity of the German navy. So the only solution seen by von Bernhardi was for Germany to arm, rearm, and arm some more. War is a necessity. War is an inevitability. There is the organic nature of the state and how it leads to struggle with others. "Every nation possesses an individuality of its own, and all progress among nations is based on their competition among themselves. As the competition among nations leads occasionally and unavoidably to differences among them, all real progress is founded upon the struggle for existence and the struggle for power prevailing among them." This is a good thing. "That struggle eliminates the weak and used-up nations, and allows strong nations possessed of a sturdy civilisation to maintain themselves and to obtain a position of predominant power until they too have fulfilled their civilising task and have to go down before young and rising nations" (26). Is this Darwinism? Were the Borborites or Phibionites, a very weird, early Gnostic sect, Christian? Making seminal fluid a central part of the communion ritual—gives an added meaning to "Take, eat; this is my body," don't ask about the blood—they were as far from Dean Farrar as von Bernhardi often seems from Charles Darwin. In the end, it is more a matter of stipulative than of lexical definition. Darwinism perhaps, but an uneasy Darwinism given the celebration of ongoing violence, an alien element from the Fatherland.

At the Point of Deluge

Pause, as the First World War is about to destroy the known world. We have taken the story down from Darwin to von Bernhardi. Are we making good on the claim that Darwinism, as we are construing it, is a religion, or more moderately, a secular religious perspective on life, in our case, on war? It has been shown that Darwin's thinking generally and his thinking specifically about war owes much to his Anglican childhood. This does not make his theory, especially his theory considered as professional science, into a religion. However, at the least, it prepares the way for a take on the main ideas—evolution through natural selection—to be given a religious gloss or even a deeper religious meaning. It is not as if we are dealing with Darwin's thinking versus the rules of soccer, for instance, even though many soccer fans do whip themselves up into something akin to religious frenzy. With Darwin's thinking, we are dealing with something covering the same ground as religion and often in the same terms. Organic characteristics, for instance, are picked out as in need of explanation, and in Darwinian thinking as in Anglican thinking the explanation involves their design-like nature. The eye is like a telescope, and for both Anglican and Darwinian that is something important and in need of explanation.

More than this, when we start to think about the sorts of things a religion does, it is clear that Darwin wants much the same from his theory—at least, he does in the *Descent*. He wants a picture that covers the deity and our thinking about its existence; he wants a story of origins, of something that gives meaning, that privileges human beings, that speaks to moral behavior, and more. Darwinism, in the sense we are understanding the term, does all of this and more. It is not given by the science strictly. As noted, in the *Descent*, Darwin is explicit that he is not disproving the existence of God through science. In talking of the beliefs of primitive people and whether the idea of a God is universal, Darwin says: "The question is of course wholly distinct from that higher one, whether there exists a Creator and Ruler of the universe; and this has been answered in the affirmative by the highest intellects that have ever lived" (Darwin 1871, 1, 65). That said, he makes it clear where he himself stands on these things. "I am aware that the assumed instinctive belief in God has been used by many persons as an argument for His existence. But this is a rash argument, as we should thus be compelled to believe in the existence of many cruel and malignant spirits, possessing only a little more power than man; for the belief in them is far more general than of a beneficent Deity." Darwin in the religious or spiritual world is a nonbeliever and—whether or not this is the only or main reason—his science eggs him on.

His science also (in his opinion) backs his belief about the special position of human beings and our moral sense toward our fellows. The science explains why we are moral. It does not explain why we should be moral. Why not be like the student Rodion Raskolnikov in Dostoevsky's *Crime and Punishment* and simply ignore or transcend morality? However, as with God, it is clear that Darwin does think we should be moral and uses his science as foundation. It is in this sense I claim that Darwinism, for Darwin, is a religion. And it is in this sense I add the claim that this extends to his treatment of war. The reasons why war is possible—likely even—come out of the nature of morality, something that has a variable reach from relatives to strangers. The nature of war and its possible decline is in line with his science, infused by his reading of the natural theology of Malthus. I repeat again that, whatever complicated knots the late nineteenth-century Anglican theologians were tying themselves in about the topic of original sin, there is certainly an Augustinian flavor to Darwin's thinking on human nature. Remember, he too had been at an English public school, seven years at an impressionable age, and at home probably getting a full blast of evangelical Christianity from his sisters, not to mention the Anglican-immersed time in Cambridge, all years before he became a Darwinian.

Darwin's successors, from Herbert Spencer on, sometimes likewise had a religious upbringing and sometimes not. There is no claim that the Anglican background of Charles Darwin was shared by or directly influential on any of them—although Calvinism of some form left its strong mark on Huxley, and I shall be picking up repeatedly on this kind of point. The successors virtually all did follow Darwin in discussing and worrying about war and relating it to their biology. Moreover, in following Darwin and his discussion of war, even if they did not always follow Darwin directly—Wallace for instance and his dislike of jingoistic pro-empire braying—they followed him in going beyond the science to claims of morality and of meaning and more. They likewise pushed in a religious direction.

If you think of the core Darwinism view as progress thanks to selection, responding to the existential void that follows on a nonexistent or indifferent God, as something that triumphs with humans at the top, we clearly see variation here. Thomas Henry Huxley is a lot less confident of progress than his erstwhile friend Herbert Spencer. Prince Kropotkin sees struggle with the elements as all-deciding in ways that others do not, and this colors his view of society and of the nature and necessity of war. Friedrich von Bernhardi seems far more committed to the ongoing existence of war than does Darwin himself, or even William James, for all that the latter seems committed to an innate, violent streak in humankind. But it is the basic picture that they are gathered around and they are joined by approaching war from this picture—one owing much to Darwinian science but going beyond anything a science can set out to do even in principle. Moreover, I would make a virtue out of the problematic. That we have the range of positions in itself points to something religion-like. We have seen the Christians with their range of views on war. Providence, yes, but Providence is various shapes and forms, some of which start to look awfully progressive.

In this, both Christianity and Darwinism point to their religious intents. Scientists differ, sometimes bitterly. We shall see this when we come to the question of whether violence in humans has always been a fact of life, as it were. Yet, for all the differences, the hope and oft-realized intention in science is that differences will be resolved. The wave theory and the particle theories of light battled it out for nearly two centuries. Then the wave theory won, and that was that—or rather, that was that until quantum mechanics came along—and then that was that in another way! Resolution is the nature of science. It is not of religion, at least not as readily. As is said about all religions, as soon as you put together two believers, you soon have three positions. That there is this controversy proves the point. Christianity is a religion. Darwinians and post-Darwinians—and I happily here include people like Vernon Kellogg because, even if they reject Darwinian theory as science, when it comes to discussions of things like war, they are working in the world created by Darwin—are in the world of religion, or at the least in a religious mode. About Friedrich von Bernhardi one can say only that, however you characterize his thinking, it is certainly not scientific. Is it philosophical? Is it religious? In the end, it doesn't really matter because for him, and his fellows, the two more or less collapse into one.

5 | Onward Christian Soldiers

THE FIRST WORLD WAR WAS, by any standard, horrendous (Keegan 1999; Stevenson 2004; Strachan 2004). Focusing on the West, in August 1914, Germany marched through Belgium and into France, hoping to strike into the heartland of the country. With incredible loss of life on both sides, the Germans were stopped in Northern France, and then the opposing nations settled into an ordeal of attrition, hoping to wear down the foe across the trenches into which both sides dug themselves (Hastings 2013). There were huge battles—in 1916, first Verdun with the Germans attacking the French and then the Somme with the British attacking the Germans—with inconclusive gains. The same was true at sea. The Battle of Jutland (fought in 1916) was the only significant encounter, and even though the Germans inflicted more damage, the British had resources enough to go on fighting with nary a blip and continue their blockade of Germany. Finally, with the coming of the Americans in 1917, for all that (in the East) Russia collapsed into revolution and the end of fighting on that front, overall the tide started to go against the Germans. There was one final desperate German effort in the spring of 1918, leading the British leader Douglas Haig to speak of "backs against the wall" (Stevenson 2011). In the end, it was of little avail, and in November 1918, an armistice was signed. "The war to end war," as it was called misleadingly—the British Prime Minister David Lloyd George added sardonically "until the next one"—was over. Above all, it was a religious war. It was a very Christian war.

The Christians Go to War

> God heard the embattled nations sing and shout
> "Gott strafe England" and "God save the King!"

God this, God that, and God the other thing –
"Good God!" said God, "I've got my work cut out!"

Just war theorists back to Augustine distinguish between the factors precipitating a war and the underlying basic causes. It is important to do this in understanding the European breakout of war that occurred at the beginning of August in 1914. The spark that lit the tinder was the assassination at the end of June, in Sarajevo, of the Archduke Ferdinand, heir to the throne in Austria. This led to an impossible-to-satisfy ultimatum by Austria toward Serbia and eventual declaration of war. In turn, Serbia's ally, Russia, started to mobilize and this in its turn brought in Germany. The military plan—the Schlieffen Plan—was to strike first at Russia's ally France in the West, going in through Belgium, and then turn to Russia. Entering Belgium brought in Britain, and the conflict was on. Italy joined in a little later on the side of the allies; Turkey was on the side of Germany and Austria. In the spring of 1917, after Germany declared unrestricted naval warfare, the United States joined the conflict along with Britain and France.

Why, after a century of peace, did it all happen? Why were so many people anticipating it? These are still all greatly contested issues. Macmillan (2014) and Clark (2014) demonstrate the difficulties of trying to make definitive judgments. No one could deny that, by the end of the nineteenth century, there was a complacency, not to say arrogance, toward one and all, by the Empire-owning and -building British. This said, a very significant factor—and one says this not in a moral or judgmental way—was Germany. Until 1871, although Prussia in the North was by far the biggest and strongest, Germany was a more-or-less loose confederation of many states, linked by a common language. Then came unification. Otto von Bismarck pulled the states together and made them into one country. Prussia became even more dominant, with Berlin as the capital of the new country and its monarch, the Kaiser, ruler of all of the land (Clark 2006). In Britain, Queen Victoria had no power of governing. As is the case today, that was for parliament. In Germany, like the Tsar in Russia, the Kaiser had significant power—as Bismarck was to discover in 1890 when he was dismissed by Wilhelm II. Reinforcing all of this was a military devotion to king and country. "I do not hear the call of my relatives, but only that of the fatherland, not the din of the fearful weapons, but only the thanks that the fatherland sends me" (Clark 2006, 221). The armed forces formed almost a rival government, with the General Staff having huge power. Starting at the end of the old century, they promoted that fierce, naval arms race with Britain, then the most powerful nation in the world.

Unified, Germany was just too big for the rest of Europe, and it felt threatened on the one side by England and France and on the other by the rapidly industrializing Russia. Hemmed in and needing to push out before it was caught forever. The biologist Peter Chalmers Mitchell had studied in Germany toward the end of the nineteenth century. He had very great respect for German life and culture. Respect tempered with realism. He sensed the walk to war, even predicting this in an article in the *Saturday Review* in 1896. "Were every German to be wiped out to-morrow, there is no English trade, no English pursuit that would not immediately expand. Were every Englishman to be wiped out to-morrow, the Germans would gain in proportion. Here is the first great racial struggle of the future. Here are two growing nations pressing against each other, man to man all over the world. One or the other has to go; one or the other will go" (reprinted in Chalmers Mitchell 1915, xxiii).[1] War had powerful causes but not necessarily good reasons. Everyone agrees that 1914 was not a 1939 situation. That was the tragedy. The Kaiser was not Hitler. Churchill, who knew the Kaiser quite well, spoke of him as amiable, and even went so far as to offer him asylum in Britain at the beginning of the Second World War. Wilhelm was weak and insecure, having a withered arm from birth, and he was proud and headstrong, but not inherently evil. He just wasn't up to the job—of preventing war or (later) of waging it. He got pushed aside by the military leaders, Paul von Hindenburg and Erich Ludendorff. By then, it was all too late.

This helps us to understand the reactions of the churchmen in both Britain and Germany, for to be quite candid this was not their finest hour. Paradoxically almost, in the decade or two before the War, Protestant leaders in England and Germany had been involved in major, ongoing projects of ecumenical friendship. Launched in 1910, "The Associated Councils of Churches in the British and German Empires for Fostering Friendly Relations between the two Peoples" had attracted the attention both of the Archbishop of Canterbury, Randall Davidson, as well as the outstanding

[1] *Rilla of Ingleside* (1921), one of the later works in the *Anne of Green Gables* series by Lucy Maud Montgomery, takes place on Prince Edward Island (Canada) during the Great War. Although the outbreak of war was unexpected, no one was surprised at the war. Everyone had long assumed that German militarism was heading Europe that way. Later editions removed the anti-German sentiments. The recently restored, original edition—including a John McCrea-type figure, who, days before death, writes the iconic poem of the whole war—is one of the most poignant fictional accounts to come from that conflict.

German spokesman for "liberal Christianity"—meaning a religion (much influenced by Schleiermacher) that worked with modern science and put a major emphasis on the already-mentioned "Social Gospel"—Adolf von Harnack. However, Anglicans are part of the establishment and the same in major respects is true of Lutherans—they look back to their founder and his backing of the authorities against the Peasants' Revolt—and so they fell naturally into looking for theological justification for their countries' actions. U-turns were executed. Overnight.

Focusing first on the Anglicans in Britain, they were in a particular quandary. On the one hand, here was their country going to war, amid huge excitement. They would not stand aside. They could not stand aside. They had sold themselves for a mess of potage. Cruel but not entirely unfair. As the established church, there were huge benefits, starting with a guaranteed presence in parliament. As the established church, there were huge costs, starting with the felt obligation to find Christian backing for the secular declaration of war by the politicians. Reasons for fighting had to be found. On the other hand, while it is not exactly true to say that Anglicanism was a religion bereft of deep theology—although one notes how quickly the greatest English theologian of the nineteenth century, John Henry Newman, scarpered off to the Church of Rome—in the years before the War, its major theological concerns were about evolution and such topics, leaving it essentially lacking resources to deal with the new, awful situation. To say the least, just war theory, the serious, traditional approach to organized conflict, was "out of fashion" (Fisher 2014, 28). The words of Mozley and Maurice were reprinted and given wide circulation, but they were really about earlier times, when Britain was painting the world red. In any case, they were infused with a confidence and optimism that few now felt. It was time for a darker theology, and a return to the church's original Augustinian foundations. No more edging up to progress and modifying Providence beyond recognition. It was time for the picture of sinful man and of the Christian obligation to fight against evil.

This highlights the differences between 1914 and 1939 and the dilemma of the church at the beginning of the former. If you are fighting Nazis, then there is little problem in saying and feeling that you are fighting against evil, or that hence one has a Christian obligation to get involved. The trouble was that the Germans of 1914 were not Nazis. They were regular folk, like us. More than one writer stressed this.

I walked in loamy Wessex lanes, afar
From rail-track and from highway, and I heard

In field and farmstead many an ancient word
Of local lineage like "Thu bist," "Er war,"
"Ich woll," "Er sholl," and by-talk similar,
Nigh as they speak who in this month's moon gird
At England's very loins, thereunto spurred
By gangs whose glory threats and slaughters are.

Then seemed a Heart crying: "Whosoever they be
At root and bottom of this, who flung this flame
Between kin folk kin tongued even as are we,
"Sinister, ugly, lurid, be their fame;
May their familiars grow to shun their name,
And their brood perish everlastingly."

("The Pity of It," written April 1915, in Hardy 1994, 500)

One obvious solution is to argue that the foe really is evil. Germans are drenched in original sin. Thanks to total propaganda insensitivity, the German High Command shoveled up tons of grist for this mill. As the German armies invaded, they allowed and encouraged the most appalling atrocities to be directed at the Belgians. Fueled with alcohol, irritated at the Belgian presumption to object and defend, civilians were executed, houses destroyed, churches desecrated. The precious university library at Louvain was burned. If this were not enough, the next year the Germans shot the English nurse, Edith Cavell.[2] Fighting a people who would do this is a Christian obligation. The jingoism is off the scale. The Right Honorable and Right Reverend Arthur F. Winnington Ingram D.D., Lord Bishop of London, was good on the breasts-cut-off-by-the-German-fiend theme (Wilkinson 1978, 94). Others, not sharing the episcopal obsessions—he was big on morality, especially sexual—focused more on the need to fight. Writing in 1914, this is Wilfred Owen—Wilfred Owen!—on the situation.

Oh meet it is and passing sweet
To live at peace with others,

[2] She was shot for helping the escape from Belgium of two hundred British and French soldiers. When I was a child, taught by two generations who had lost so much, along with sixteen-year-old Jack Cornwell—awarded the Victoria Cross, posthumously, for his gallantry at the Battle of Jutland (1916)—the heroism of Nurse Cavell was constantly before us. To this day, I cannot pass her statue in Trafalgar Square without tears welling. In fact, for all that her famous last words were "patriotism is not enough, I must have no hate in my heart," the Germans did have a point. They were just so monumentally stupid. Could they not predict what would happen and how their act would be used against them? Nurse Cavell was driven by her strong Anglican faith. Lest at times this book seems unduly harsh on the Anglicans, be it noted that I see that full stories are always far more complex.

But sweeter still and far more meet
To die in war for brothers.

Parenthetically, one might add that there was a strong element of hypocrisy here in all of this frenzied patriotism. Up to this point, the British loathed and detested the Belgians: with reason, given the latter's appalling practices toward Africans in the Congo. Joseph Conrad's *The Heart of Darkness* dates from 1899.

In parallel, the other, no-less-obvious solution was to turn the war into a religious war. When you are fighting the infidel, generally there is no great presumption that the individual people you are fighting are all that sinful. Indeed, you might have great respect for them as warriors and a secret regret that you too cannot have four wives. However, if you start to argue that being in itself an infidel is sinful—refusal to accept Jesus as your savior—then you have a ready Christian obligation to go to war. Although for some years he wrestled with this problem, ultimately Augustine was categorical. Religious war is justified. Translated into the present: Germans are infidels! The Anglican Church was up to the challenge of proving this point, starting with the fact that Germans are not English. Shakespeare is our dramatist and poet. Far better than that effete Goethe chappie with his unspeakable obsessions about women's dirty underwear. You pump up the differences between the absurd Kaiser and our own dear royal family—who incidentally found it politic in 1917 to change its name from the house of Saxe-Coburg and Gotha to the house of Windsor. You tar the others with the same brush. For instance, the novelist John Buchan in his novel *Greenmantle*, published in 1916 when he was working for the British War Propaganda Bureau—he was to become Director of Information—introduced Colonel Ulrich von Stumm, a supposedly typical German officer.

> He was a perfect mountain of a fellow, six and a half feet if he was an inch, with shoulders on him like a shorthorn bull. He was in uniform and the black-and-white ribbon of the Iron Cross showed at a buttonhole. His tunic was all wrinkled and strained as if it could scarcely contain his huge chest, and mighty hands were clasped over his stomach. That man must have had the length of reach of a gorilla. He had a great, lazy, smiling face, with a square cleft chin which stuck out beyond the rest. His brow retreated and the stubby back of his head ran forward to meet it, while his neck below bulged out over his collar. His head was exactly the shape of a pear with the sharp end topmost. ([1916] 1992, 146)

With a man like this, one is almost primed to learn that, in his private quarters, Colonel von Stumm has a taste in knickknacks and dainty furniture, and there are few things he has yet to learn about embroidery. Fighting people like this is a Christian obligation.

A Religious War

The Great War was a religious war! The opposing side was the anti-Christ, and justifiably one could oppose and fight against it. "Not just incidentally but repeatedly and centrally, official statements and propaganda declare that the war is being fought for God's cause, or for his glory, and such claims pervade the media and organs of popular culture." Furthermore, "they identify the state and its armed forces as agents or instruments of God. Advancing the nation's cause and interests is indistinguishable from promoting and defending God's cause or (in a Christian context) of bringing his kingdom on earth" (Jenkins 2014, 6). Even if they had been that way inclined, for the British "Just War" justification was pushed to one side in favor of an "Apocalyptic Crusade." A move that never stopped giving for at once—even though thoughts of progress were now gone—one could still pick up on the moral improvement open to the Christians. A reason why Mozley and Maurice were much liked was how this kind of message could be preserved in and reinforced by war. Maurice spoke of war leading to "wholesome family life" rather than "corrupt family life."

One sees this cherishing of the ethical mixed up with the healthy at once in the sermons of the British clergy when war broke upon the nation in 1914. Thus Winnington Ingram in St. Paul's Cathedral on August 9. "It may be the best lesson possible for the well-to-do to endure 'eternal hardness' as good soldiers of CHRIST JESUS, and the best lesson in brotherhood we shall ever have to endure it with the poor side by side. May it not be that this cup of hardship which we drink together will turn out to be the very draught which we need?" (Winnington Ingram 1914, 7). I teased William James about his prescriptions for the young. There is a Germanic steeliness about Winnington Ingram that does not encourage humor. "Has there not crept a softness over the nation, a passion for amusement, a love of luxury among the rich, and of mere physical comfort among the middle class? Not such was the nation which made the Empire, which crushed the Armada, which braved the hardships of old, and drove the English 'hearts of oak seaward round the world.' We believe that the old spirit is here just the same, but it needed a purifying, cleansing draught to bring it back to its old strength and purity again; and for that second reason, the cup which our FATHER has given us, shall we not drink it?" (7). War is needed for—personal and societal health, not to mention religious observance.

More generally, the Anglican clergy simply became the cheerleaders of the war effort—Germans evil, war necessary. William Temple, son of one Archbishop of Canterbury and later become himself another Archbishop of Canterbury, is rightly remembered for his strong social conscience and in the Second World War for his stand on behalf of the Jews. Until 1914, he was headmaster of Repton—a venerable English public school—a traditional career position for a man on the way to the higher echelons of the Anglican Church. He preached a full-blooded call to arms. Unlike the Quakers and unlike Tolstoy, Jesus was no pacifist. "The watchword of this doctrine is 'Non-resistance,' 'resist not evil.' But we must interpret Christ's teaching by his actions. He did not resist physically; but He did resist the evil of His day, even to death. He did not say: 'The people of Jerusalem are very obstinate; here in Galilee are crowds waiting eagerly for the Gospel; here then we will work in peace.' On the contrary, 'He set His face steadfastly to go up to Jerusalem.' He did not stay out of the fight; He went into it" (Temple 1914, 6–7). Of course, there are rules of conduct in war. Would that the Germans remembered these. "There can, I suppose, be no doubt at all that German soldiers have been guilty of atrocities" (8).

Temple did not forget that he was a man of God. Above all, the Christian will pray for his enemies. "Either we must rewrite the old words to run, 'Do good to them that hate you, after you have taught them a good lesson; pray for them that persecute you, when they are wounded'—or else we must change our whole attitude to our enemies, alike in action and in prayer" (9–10). Temple also recognized in good Augustinian fashion that war is evil. In equally good Augustinian fashion, sometimes it must be fought. " 'Wretched man that I am, who shall deliver me from the body of this death?' For it is the hideous result of sui that it brings us into a choice where even the rightest thing that we can do is something evil; the choice is between the greater and the lesser evil. And though we are right, and absolutely right, in choosing the lesser evil, it is still evil, for it is still not perfect obedience to the holy will of God" (13). Note in strictly Augustinian terms, fighting in itself is a bad thing, but sometimes it is morally right for us to fight.

These sorts of moral and theological niceties did not trouble the truly dreadful Winnington Ingram. This, from a sermon in 1915.

> To save the freedom of the world, to save Liberty's own self, to save the honour of women and the innocence of children, everything that is noblest in Europe, everyone that loves freedom and honour, everyone that puts principle above ease, and life itself beyond mere living, are banded in a great crusade—we cannot deny it—to kill Germans: to kill them, not for the sake of killing, but to save the world; to kill the good as well as the bad, to kill the young men as well as the old, to kill those who have shown kindness to our wounded as well as those fiends who crucified the Canadian sergeant, who superintended the Armenian massacres, who sank the *Lusitania*, and who turned the machine-guns on the civilians of Aerschott and Louvain—and to kill them lest the civilisation of the world should itself be killed. (Quoted in Marrin 1974, 175)

What of the men out there killing and dying? Did they have no theological stake in matters? This poem was written at the end of June 1916, by the son of the (Anglican) Bishop of Saint Edmundsbury and Ipswich, Lieutenant Noel Hodgson, brilliant classics scholar and sportsman, known as "Smiler" to his friends. The reference in the first part of the verse is to the view from the hill on which sits Durham Cathedral, in the town where Hodgson went to school.

> I, that on my familiar hill
> Saw with uncomprehending eyes
> A hundred of thy sunsets spill
> Their fresh and sanguine sacrifice,
> Ere the sun swings his noonday sword
> Must say goodbye to all of this;
> By all delights that I shall miss,
> Help me to die, O Lord.

Hodgson's God did not fail him. On the first day of the Battle of the Somme, July 1, 1916, thanks to a bullet through his neck, he died instantly. He is buried in Grave 3, Row "A" at Devonshire Cemetery in Mansell Copse. What the American preacher Harry Fosdick, quoting the Old Testament, was to call a blood sacrifice offered "without blemish before the Lord." Would that the Lord had not been quite so demanding. On that same day that Hodgson died, so did almost twenty thousand more of his fellow countrymen (Philpott 2009). In total, through the battle, over four hundred thousand casualties from Britain and its Empire, over two hundred thousand French, and probably as many Germans as the two combined. At Beaumont-Hamel, France, 780 officers and men of the Royal Newfoundland Regiment went over the top of the trench. The next morning, only 68 men answered the roll call. Nearly a quarter of the then-dominion's young men never returned. To this day, when on July 1 the rest of Canada commemorates the founding of the country, Newfoundland (which did not join Confederation until 1949) pauses in sorry memory.

One does sense a melancholy. As one should. What had started almost as a jolly jape out of the *Boy's Own Paper* had descended into a nightmare calling for Hieronymus Bosch. Wilfred Owen was writing very different poetry now. How could Temple sleep at night, spending his days urging the young men of the nation on to war, even as knew that he himself would never have to make the supreme sacrifice? The school's memorial wall commemorates 355 Repton boys who died in the Great War. From one school alone. 355! Even Winnington Ingram knew there was something dreadfully wrong. Always there is that haunting reflection that our enemies are not alien beings. They are our fellow humans, our kin. Literally in the case of the royal families. Kaiser Wilhelm and George V were both grandsons of Queen Victoria. John Buchan—he was a Presbyterian but doing the Anglicans' work—has the hero of *Greenmantle*, Richard Hannay, somewhat improbably knock Colonel von Stumm senseless and escape. He travels incognito through Germany, falls sick, and is nursed by a woman with small children and a husband absent at war. On leaving, he reflects.

> I used to want to see the whole land of the Boche given up to fire and sword. I thought we could never end the war properly without giving the Huns some of their own medicine. But that woodcutter's cottage cured me of such nightmares. I was for punishing the guilty but letting the innocent go free. It was our business to thank God and keep our hands clean from the ugly blunders to which Germany's madness had driven her. What good would it do Christian folk to burn poor little huts like this and leave children's bodies by the wayside? To be able to laugh and to be merciful are the only things that make man better than the beasts. (Buchan 1916, 191)

Winnington Ingram of all people was German speaking and a year or two before had said: "We are cousins, nay, we are brothers to the Germans, and therefore, as a nation, I say, '*Wir alle lieben Deutschland und die Deutschen*'" (German Christian Churches 1908, 188). These men were dreadful. They were appalling. More truly, they were out of their depth. They were essentially decent, faced with a situation of which they had no experience, armed only with an ideology that simply was not adequate, compromised by their church's symbiotic connection to the state. Few of us are simply all black or all white. Sickened by the carnage as the war dragged on, Winnington Ingram spoke strongly against the bombing of German cities. In mid-June 1917, in an air raid, the Germans killed over one hundred people, including many children in London. The *Times* reported the bishop as saying that he "did not believe that the mourners should wish that 16 German babies should lie dead to avenge their dead" (Bell 2012, 43). This was not a belief shared by the editor of *The Modern Churchman*: "If the only way to protect adequately an English babe is to kill a German babe, then it is the duty of our authorities, however repugnant, to do it" (Wilkinson 1978, 101–102).

German Christianity

Perhaps one should not be too harsh on the British. Apart from anything else, most of the Anglican spokesmen were bishops and they were pastoral figures, not trained academics. From my experience, to be the headmaster of a public school does not demand great scholarly abilities, nor does it inculcate more than blinkered moral awareness. Farrar was not the only head who thought that illicit production of seminal fluid was the true topic of Plato's *Republic*, Aristotle's *Nicomachean Ethics*, and St. Thomas's *Summa Theologica*, not to mention the underlying theme of all of Shakespeare's plays. Think Hamlet. One expects more of Germans, home of the leading theologians and philosophers of the nineteenth century, but one is thereby even more greatly disappointed. All too typical on their side was a Lutheran pastor preaching in the early days of the war: "German Christianity represents the right relation between Christ and His disciples, and our nature the most perfect consummation of Christianity as a whole. We fight, then, not only for our land and our people; no, for humanity in its most mature form of development; *in a word, for Christianity as against degeneration and barbarism*" (Bang 1917, 69). Although there was something of a reluctance to admit that Germany was in any sense inadequate, let alone in the wrong, it was admitted that a certain softness had crept in and remedial actions were appropriate. "We had squandered our best forces and desires upon perishable things, the temporal had become our all-powerful idol, the eternal had vanished from our thoughts. We had entirely lost ourselves in mechanical realities, and forfeited all insight into the importance of spiritual values" (108).

The leading churchmen were little better (Hoover 1989; Jenkins 2014). Adolf von Harnack, always close to the Prussian authorities—he was the first president of the newly founded Kaiser Wilhelm Institute—had long argued strongly for the legitimacy of a state's standing army ("Supreme God . . . by thee we are victorious and happy"), and when the war came he was a signatory of the notorious "Manifesto of the Ninety-Three." This was a defense, by leading intellectuals, of the German behavior—with reason usually taken to be morally atrocious—in the early months in Belgium. "*It is not true* that Germany is guilty of having caused this war. Neither the people, the Government, nor the Kaiser wanted war. Germany did her utmost to prevent it; for this assertion the world has documental proof. Often enough during the twenty-six years of his reign has Wilhelm II shown himself to be the upholder of peace, and often enough has this fact been acknowledged by our opponents" (Professors of Germany 1914, 284).

Even Rudolf Steiner, the Austrian clairvoyant—known today chiefly as the founder of the Waldorf school system and of biodynamic agriculture—got into the business, in 1914 writing poetry to the then Chief of the General Staff of the German army, Helmuth von Moltke:

> Victorious will be the power,
> Which by the fate of times,
> Is predestined for that people,
> Who, protected by the spirit,
> Will, for mankind's salvation,
> In Europe's heart,
> Wrest light from the battle.
>
> (Meyer 1997, 134)

And so on and so forth. Von Moltke lost his job a day or two later for losing the Battle of the Marne (when the French stopped the German advance) and died in 1916. Fortunately, that did not stop Steiner from continuing their conversation, reporting back that that the Field Marshall was not pleased at the way things had gone. In 1919, sadly, the departed soldier reflected: "It was Germany's destiny not to murder itself but to be murdered" (233).

America Joins In

We have learned how much of this thinking and behavior (on both sides) is a function of the tight connections in Europe between state and church;

but things were much the same in America, where there is separation of church and state. In the name of Jesus, Congregationalist minister Newell Dwight Hillis thought that, after the war is over, we might have to sterilize ten million German men and segregate off their women—"that when this generation of German goes, civilized cities, states and races may be rid of this awful cancer that must be cut clean out of the body of society" (Hillis 1918, 59). Lyman Abbott, Congregationalist minister, editor, writer, representative of liberal theology and social reform, enthusiast for evolution, was another who saw everything good in America and her allies and nothing good coming from the Germanic heartland of Europe. As slavery has been overthrown in the last century so the vile machinations of Germany will be thrown over in this century. In the words of "one of England's greatest poets" (Tennyson):

Form! form! Riflemen form!
Ready, be ready to meet the storm!
Riflemen, riflemen, riflemen form!

(Abbott 1918, 9)

Expectedly the health-giving nature of conflict got a special mention. "For it cannot be doubted that Americans were growing soft, easy, adipose. Our prosperity was poisoning us. We were fast assuming the fatal falsehood that happiness is the end of life" (54–55). Even language was suspect and indicative of moral decay. "Our current phrases, "A happy New Year," "Many happy returns of the day," "A long and happy life to you," were conventional, but they expressed what was becoming a dominant desire for ourselves and for our friends." Safety first "generally meant comfort first" (55). Now, thanks to war: "Character, wisdom, righteousness, education, human development, progress, growth, greatness of heart and greatness of mind are some of the phrases which have been employed. The word I like best is the one which is most frequent in the New Testament: Life." Do not think we are alone in this struggle. "I can see that God is doing something for us that is much better than stopping the war; he is inspiring us with courage to win it" (51–52).

Pacifism

Was there no one prepared to take a Christian stand against this sort of thing? Was pacifism totally washed away by this bellicose militarism? There was much gory hypothesizing about the consequences of pacifism, usually involving lust-driven German soldiers who, before the defenseless Englishman, "strip naked and rape in his presence his mother, his wife, his sister, and his daughter—or indeed any Englishwoman—one man following another in succession" and then, as a kind of encore, proceed "to slice off the breasts of these women, hack off their limbs, disembowel them by hacking them open, or to kill them in other ways" (Solano 1918, 50). Interesting is the extent to which the clergy tended to equate pacifism with precisely that moral flabbiness and lack of health that war is able to remedy. "Not every peace has moral value. A peace that serves only to increase material culture and sensual enjoyment, that weakens the spiritual life and leads nations into practical materialism would be thoroughly bad" (Titius 1915, 3–4, translation in Hoover 1989, 111).

"Sensual enjoyment"! One wonders how many Quakers this particular (German) clergyman had known. Paradoxically, he who was the best-known figure to oppose the war was he who, before Richard Dawkins, was Britain's best-known atheist. By the time of war's outbreak, Bertrand Russell—aristocrat (his grandfather was Prime Minister), brilliant mathematician and philosopher, and well known for his contempt of conventional mores (especially sexual)—was too old for active service. This did not stop him lecturing non-stop against the conflict until he managed to get himself locked up, where he promptly took the opportunity to write a book on the philosophy of mathematics. Russell thought that the behavior of the Germans had been appalling, but he was not much more keen on the behavior of Britain, especially in its refusal in prewar years to assuage some of the worries and insecurities of the Germans. War itself he thought to be simple madness. "The utmost evil that the enemy could inflict through an unfavourable peace would be a trifle compared to the evil which all the nations inflict upon themselves by continuing to fight" (Russell 1916, 84). In light of the discussion of secular thinking to come in the next chapter, already noted is that Russell was no big friend of Darwinism, at least as applied to philosophical issues like these (Cunningham 1996). Not only did he write strongly against those keen Darwinians, the American Pragmatists, but he followed Moore in disliking any evolutionary approach to philosophy, positively because back then he was a neo-Platonist and negatively predicated on a dislike of Herbert Spencer's too-often naïve eclecticism. More than the evolutionism, Russell was against Spencer's embrace of the monistic thought characteristic of German thinking, which tends to such beliefs as that the state is an organism. He was ever for the individual.

Russell was not alone in his opposition to the war, and there were those for whom the Sermon on the Mount was definitive. Christians grouped around the nondenominational "Fellowship of Reconciliation." Aggressively pacifist—"as Christians, we are forbidden to wage war," hence, "our loyalty to our country, to humanity, to the Church Universal, and to Jesus Christ, our Lord and Master, calls us instead to a life of service for the Enthronement of Love" (Barrett 2014, 48)—there is the flavor of the eschatological underpinnings so common in Christian repudiations of war. It is not so much that pacifism leads to happy consequences, although there were those prepared to argue that case, but that the command to turn the other cheek is absolute. The Fellowship stood in an old tradition. One gets always the sense of a Christianity focused on the Jesus of the Beatitudes, one that makes God's dictate totally binding and our free choice absolute, rather than a Christianity focused on the Paul of the Acts and the Epistles, one that leads to (let us say) a more nuanced reading of the words of Jesus and of course makes our choice constrained by the inherited taint of Adam's sin. These pacifist Christians rejected just war theory because they were part of (or attracted to) a different tradition in their faith rather than, like the warmongering bishops, because they were ignorant of or indifferent to just war theory. We have seen this kind of non-Augustinian Christianity in the thinking of Immanuel Kant.

Prominent Anglicans associated with the Fellowship of Reconciliation included the socialist politician George Landsbury, later to be leader of the parliamentary Labour Party. Through the War, he was the editor of the *Daily Herald*, opposed to the conflict and then a supporter of the Russian Revolution. This points to the fact that although his Anglican faith was deep and important, Landsbury's opposition to war was based also on his socialist beliefs that the burdens of such conflicts rest always more on the backs of the lower, working classes. Maude Royden, High Anglican and daughter of a rich shipping magnate, stood firm against war. She was not so much unmoved by the German atrocities in Belgium, but convinced that the right response was neutrality, an almost aggressive refusal to take up arms. Against those who thought otherwise: "I, too, would have risked something—everything, indeed—to win, not a devastated and ruined Belgium, but Belgium unscathed." Showing the end-times setting of her thinking: "For such a nation could not die, although for nations as for individuals, it is true that sometimes they must lose their lives to save them" (Barrett 2014, 75). Royden, in spirit (and often in practice) a female Anglican priest before her time, was to show great bravery when attacked by a mob. She and her fellows sat passively and, although seized at one point by the throat, survived unharmed. Bernard Walke was an even-more-committed Anglo-Catholic, like his father and grandfather before him a priest in the Tractarian tradition (that which had fostered Newman), and as ardent a pacifist as Landsbury and Royden. He insisted on visiting incarcerated pacifists, even preaching to them, a practice that earned the scorn of the crazy, would-be Messiahs who thought that in their persecution they were enacting the time on the Cross, but appreciated nevertheless by grave Quakers, whose absolutist convictions had brought them behind bars.

In going to jail, Russell was the exception rather than the norm—in Britain at least. Out of rather more than fifteen thousand who declared themselves "conscientious objectors" to war, only about a thousand ended up inside. This was in part a function of the fact (already noted) that many pacifists, notably Quakers, felt that they could support the war effort in some sense, so long as they were not actually fighting. They joined ambulance services and the like and, to their great credit, many proved to be very brave people indeed. In his thriller, *Mr Standfast*, John Buchan gives a very sympathetic portrayal of one such conscientious objector. The absolute war opponents were often from tiny apocalyptic sects like the already-mentioned Jehovah's Witnesses, who refused entirely to participate in any state-run activities, expecting that the End of the World is nigh, anyway. Sometimes the state-sponsored retaliation puts one in mind of Germans in Belgium, with people dragged to France for courts martial. Fortunately, wiser political heads prevailed and the military were forbidden to execute "conchies."

In America, the Episcopalian pacifist bishop of Utah, Paul Jones, got the sack. "It seems abundantly manifest that an end has come to the usefulness of the Bishop in his present field, and no earnestness of effort on his part would suffice to regain it" (Barrett 2014, 102.) It is nice to be able to report that, in 1998, the bishop was added to the church's Calendar of Saints. His feast day is September 4, the anniversary of his death. No such honor has been extended to the Catholic Ben Salmon, another who protested, against both his state and his church. In 1917, to the president, Woodrow Wilson, he wrote: "Regardless of nationality, all men are brothers. God is "our Father who art in heaven." The commandment "Thou shalt not kill" is unconditional and inexorable." So do what you will with me. "When human law conflicts with Divine law, my duty is clear. Conscience, my infallible guide, impels me to tell you that prison, death, or both, are infinitely preferable to joining any branch of the Army" (Goff 2015, 292). For all that the army persecuted him, at one point sentencing him to death, Salmon did not budge. With the country's Catholic leaders virtually saying that to refuse conscription was close to treason, he got little sympathy there either. In prison, Salmon was refused communion.

At least part of the episcopal pharisaical behavior would have stemmed from the general insecurity of usually lower-class and little-respected Catholics—Irish, Italian, Spanish—in a Protestant country. It was hardly just that. Everyone was into war support and boosterism. The Archbishop of Canterbury assured his listeners that the Muslims had little to teach the Church of England about the immediate post-death rewards of fallen warriors. "It is the function of Christianity, of us would-be followers of Christ, to raise this unacknowledging trustfulness and self-giving out of dumb sub-consciousness and to give it speech and to crown it with the glory of the fully human self-devotion." So buck up and get on with it. "His will to be done—and that to give him joy is the supreme end of man" (Davidson 1919, 107).

The Germans were on board with this. "Truly, here is holy ground. Here is the gate of heaven. Sacrifice is the key that breaks it open" (Hoover 1989, 109).

Conclusion

Any summing up of the role in the Great War of Christianity—at least the role of the ministers of Christianity—can only restate the obvious. Looking back, a hundred years on, one is amazed that so many fought and suffered and died for causes that truly do not seem worth the candle. We have already made the salient points. There was no Hitler. No Nazi regime striking out across Europe to conquer and to displace. No vile ideologies of oppression that should sicken any decent-thinking person. These were cousins killing cousins. And yet people did fight and they did fight on, to the death. A major reason is simply that they were urged to do this by people speaking in the name of Jesus Christ. This is true for both sides. The Germans were particularly militant, but the British and then the Americans were hardly less so. This continued right to the bloody and bitter end, and even beyond. In a Christmas-day sermon in 1918, not two months since the Armistice brought an end to the slaughter, Pastor Gerhard Tolzien preached: "Germany and Christianity will again harmonize! Germanness and Christianity have already harmonized well. May our German-Christian festival testify to that. Germanness and Christianity also harmonize in our one-hundred-year old song. And one hundred years from now nothing will be lost or spoiled" (Hoover 1989, 122).

"The Church, to an unfortunate degree, had become an instrument of the State and in too many pulpits the preacher had assumed the role of a recruiting sergeant." These are the sad, sad words of an eminent Scottish clergyman, looking back over the span of four decades.

> I said many things from my pulpit during the first six months of my ministry that I deeply regret. It is no excuse to say that many preachers were doing the same thing. I still feel ashamed when I recall on one occasion—about the time of Haig's "Our backs are to the walls" message—that anyone who talked of initiating peace negotiations with the rulers of Germany was a moral and spiritual leper who ought to be shunned and cut by every decent-minded and honest man! (Warr 1960, 118–119)

Decent men, caught in a situation of which they had no experience, armed only with an ideology that was not adequate.

> "Peace upon earth!" was said. We sing it,
> And pay a million priests to bring it.
> After two thousand years of mass
> We've got as far as poison-gas.
>
> ("Christmas 1924," in Hardy 1994, 848)

Ultimately, although no appeals were made to just war theory, everything (for the non-pacifists) was set in an Augustinian framework. We humans are sinners. War will occur. We must sometimes fight in a just cause. Hopes of progress are chimerical and dangerous. Humankind will ever be humankind. Providence alone can change this. Unfortunately, the moral theological and philosophical consequences of this thinking were swamped or colored by the political needs of the day. They did not have to be. Apart from that small group of pacifists, who were thinking in other directions, there was one major Christian voice, firmly in the Augustinian tradition and as loud and clear against this particular war. This was the newly elected (September 1914) Pope Benedict XV. He was unambiguous, calling the war the "suicide of Europe." Among the many of his condemnations was his *Apostolic Exhortation: To the Peoples now at War and to Their Rulers* of 1915. "In the holy name of God, in the name of our heavenly Father and Lord, by the Blessed Blood of Jesus, price of man's redemption, We conjure You, whom Divine

Providence has placed over the Nations at war, to put an end at last to this horrible slaughter, which for a whole year has dishonoured Europe. It is the blood of brothers that is being poured out on land and sea." Continuing: "You bear before God and man the tremendous responsibility of peace and war; give ear to Our prayer, to the fatherly voice of the Vicar of the Eternal and Supreme Judge, to Whom you must render an account as well of your public undertakings, as of your own individual deeds" (Benedict XV 1915, 435).

This cut little ice. The Germans, run by Prussian Protestants, had little time for one who headed the Roman Catholic Church. The French, coreligionists, distrusted anyone who did not simply inveigh against the Germans. The British? Well, the Pope was a foreigner after all. Not the first or the last time that untold damage has come from that kind of thinking.

The Bishop tells us: 'When the boys come back
'They will not be the same; for they'll have fought
'In a just cause: they lead the last attack
'On Anti-Christ; their comrades' blood has bought
'New right to breed an honourable race,
'They have challenged Death and dared him face to face.'

'We're none of us the same!' the boys reply.
'For George lost both his legs; and Bill's stone blind;
'Poor Jim's shot through the lungs and like to die;
'And Bert's gone syphilitic: you'll not find
'A chap who's served that hasn't found some change.'
And the Bishop said: 'The ways of God are strange!'

("They" by Siegfried Sassoon 1917)

6 | The Biology of War

TOWARD THE END OF THE nineteenth century, a thirst for education followed the upward rise of America. It was then that some of the truly great universities of the world—Johns Hopkins, Stanford, Chicago, and others—were founded and began to flourish. It was the newly unified Germany that set the pace for higher education in the later years of the nineteenth century, so it is no surprise that its ideas made their way West into these new institutions of higher learning.

Vernon Kellogg

We have met already one of the first biology professors at Stanford, the Kansas-born entomologist Vernon Lyman Kellogg. He stayed until the Great War, when he joined his former student Herbert Hoover, the future president, doing war relief work in Belgium. He was fitted for this for he already knew Germany and Germans at first hand, having (like so many of his generation) done graduate work in that land now at war with Britain and France. This showed its influence—as we learned, although Kellogg was ever an ardent evolutionist, he was as ever doubtful about Darwinian selection as an adequate mechanism of change. Going back to the Romantic era and transcendentalist philosophy of Goethe and Schelling and others, the emphasis in Germany had always been on form over function (Richards 2003). The interest had been on similarities, homologies, between organisms rather than on the design-like uses of organic features, those very design-like uses that natural selection sets out to find and understand (Russell 1916; Ruse 2003). Thanks particularly to Agassiz who (as noted) was immersed in this tradition, American professional evolutionism had an odor of transcendentalism. For one such as Kellogg—thanks to influences both at home and abroad—the wonder would have been had he not doubted the power of natural selection.

In *Darwinism Today,* remember, the sad conclusion is that natural selection will not do. "Darwinism, as the all-sufficient or even most important causo-mechanical factor in species-forming and hence as the sufficient explanation of descent, is discredited and cast down." This did not discourage Kellogg from hammering away at things. Just before the Great War, he followed up his general overview of evolution with a smaller book on human evolution. Kellogg was religious; raised a Methodist; not at all fanatically Christian; and in the words of his biographer, eager "to defuse the conflict over evolution by persuading the public that evolution and religion were completely compatible with one another" (Mark Largent, unpublished). Officially, Methodists accept the Augustinian line on original sin, but although he was certainly convinced of the sinfulness of humankind, one never senses that Kellogg was going to be constrained by the theology of nigh two millennia past. Apparently, the argument from design played a large role in his thinking. "In outlining his claim that a scientist can easily be a believer in God, Kellogg drew from his experience as a naturalist, which he said revealed to him 'much of the admirable order in Nature' and emphasized the role of a divine creator in both the matter and the natural processes that surrounded us" (Mark Largent, unpublished).

Whether stemming directly from his religion or not, Kellogg was a committed pacifist who saw war as dropping away as unnecessary, as humans develop their more altruistic side, helping others and reaching out more and more broadly. For all that in *Darwinism Today* he had professed ignorance of mechanisms, some kind of Lamarckian process seems to have been the fuel of change. With a fondness for capitalizing nouns showing that it was not just biological ideas that Kellogg took from Germany, he wrote: "just as Evolution made him, with his need, a Fighter, and taught him War, so now, with the passing of this need, with the substitution of reason and altruism for instinct and egoism, Evolution will make him a Man of peace and goodwill, and will take War from him" (Kellogg 1912, 140–141). Although, as Darwin himself showed, the English were good at this game, one regrets to say that Kellogg inherited an all-too-familiar view of racial superiority—"War is already an anachronism in the life of Homo sapiens. The evolutionary mode of the Blond race has moved beyond it. The leaders will fall in the mode or fall out of their plans. Homo superioris will be, whatever else he is, BEYOND WAR" (172).

War as Eugenically Stupid

Right at the end of his book, Kellogg showed that there is some life in natural selection yet, if its presence is denied and if it is invoked in a eugenical context. "War, to the biologist seems, above all else, stupid. It is so racially dangerous. It so flies in the face of all that makes for human evolutionary advance, and is so utterly without shadow of serious scientific reason for its maintenance." Somewhat mysteriously, Kellogg then adds: "it is not natural selection in Man, nor in any way the counterpart of it" (Kellogg 1912, 168). One would have thought, in the context, that this is precisely what natural selection is, but clearly Kellogg is so committed to an upward picture of life's progress, an ongoing force in a very Germanic fashion, he cannot see that natural selection as such can be anything for the good. A rose by another name, apparently.

This idea that war is dysfunctional, going against the evolutionary grain, was very popular in the years leading up to the First World War. Kellogg's chief and mentor, David Starr Jordan, the biologist president of Stanford University, was much given to reflections and speeches on the topic. "Those who fall in war are the young men of the nations, the men between the ages of eighteen and thirty-five, without blemish so far as may be—the men of courage, alertness, dash and recklessness, the men who value their lives as nought in the service of the nation." Those who are left behind are the dregs, although unfortunately it is they who will now make the running. "The man who is left, is for better and for worse the reverse of all this, and it is he who determines what the future of the nation shall be." Alas and alack: "the angle of divergence between what might have been and what has been, will be determined by the percentage of strong men slain on the field of glory" (Jordan 1907, 72).

There is the dreadful tale of France after the Napoleonic wars. "The unfailing result of this must be the failure in the nation of those qualities most sought in the soldier. The result is a crippled nation, 'Une nation blessee,' to use the words of an honoured professor in the University of Paris." Not that things are much better across the Channel. All of those wars of conquest have taken their toll. In parish church after parish church, in cathedral after cathedral, one sees the monuments to young men who fell for the glory of empire. "What would be the effect on England if all of these 'unreturning brave' and all that should have been their descendants could be numbered among her sons today?" (86). Of course, as we learned from Dean Farrar, not everyone who fell was a stellar example of a human being. "But most of them were worthy. Most of them were brave and true, and most of them looked out on life with 'frank blue British eyes'" (88). One hates to ask about the private lives of the brown-eyed losers. Non-stop headaches one suspects.

Jordan kept up the pace, just before the war lecturing in England to an audience including her grace the Duchess of Marlborough and Major Leonard Darwin, youngest son of Charles Darwin and president of the Eugenics Society. The farmyard shows the way. "If we breed from the weak or feebleminded and vicious, those, in a word, which are undesirable in one way or another, you will have the same general type: 'Like the seed is the harvest'" (Jordan 1913, 199). Interestingly the effects of war on Germany are not as bad as one might expect. "The evidence of the havoc of war is not so clear in Germany as in most other lands of Europe. Perhaps massacre and desolation destroyed the weak as often as the strong" (206). Given that Jordan was an appalling racist, it could be that his feelings about eugenics were battling with his feelings about racial superiority. Either way, war is no good thing. "The same motive, the same lesson, lasts through all ages, and it finds keen expression in the words of the wisest man of our early national history, Benjamin Franklin, 'Wars are not paid for in war time: the bill comes later'" (213).

Kellogg got on this bandwagon in a big way. Whether or not it should be called "natural selection," the evolutionary effects of war are just awful. Race deterioration is nigh inevitable. Kellogg noted that "three points, all of which go to substantiate this claim, are particularly brought out: first, the conditions of the formation of armies (selection of soldiers); second, a case of actual, measurable, physical, racial deterioration caused by excessive militarism; third, the conspicuous association with militarism of certain race-deteriorating disease" (Kellogg 1916, 199). This matter of sexually transmitted diseases was something of an obsession with Kellogg. "Venereal disease finds in armies a veritable breeding-ground. That such disease is highly dysgenic, i.e. race-deteriorating in influence, is indisputable. The frightful effects of syphilis, and its direct communication from parents to children, are fairly well known popularly" (196). Given the way that, as a Lamarckian, Kellogg rather mixed up the effects of the environment and the powers of heredity, it is small surprise that he was prepared to regard syphilis as on a par with, indeed part of, more obvious hereditable diseases like gross physical ailments. "Syphilis is the hereditary disease par excellence. Its hereditary effects are more inevitable, more multiple, more diverse, and more disastrous in their results on the progeny and the race than in the case of any other disease. Syphilis, in fact, has a more harmful influence on the species than on the individual" (197).

His readers agreed with Kellogg. Although not all biologists, including those keen on eugenics, accepted everything at face value. Ronald A. Fisher, ardent eugenicist and on his way to becoming one of the most influential Darwinian evolutionists of the twentieth century, bought into some of Kellogg's arguments, but not all. "There is apparently no question of the eugenic significance of the deaths of soldiers in battle. The soldier is selected for valuable physical qualities, and in the case of voluntary enlistment, for courage and devotion to duty also; he is unquestionably on the average superior to the civilian, and his death by increasing the proportion of the latter is an injury to the race" (Fisher 1916, 264). Fisher was less sure about diseases and the like. "Of the deaths of soldiers by sickness during a campaign it is less easy to judge. Its effect is undoubtedly that the veteran is more hardy and possessed of better innate qualities than the recruit; but in the absence of data it is impossible to tell whether this effect exceeds, or falls short of the injury done by the reduction of the proportion of soldiers to rejects" (264).

Fisher was unimpressed by Kellogg's thinking about venereal disease. "To compare, as Mr. Kellogg repeatedly does in the British statistics, the proportion of recruits rejected on account of syphilis with the proportion found in the army is grossly and inexcusably misleading. Apart from the fact that a syphilitic is unlikely to offer himself for a medical examination, the recruits are mostly boys of 18 or 19, and it is just in the succeeding ten years that this disease and gonorrhea find most of their victims" (265). One suspects that this is a time when Fisher and his science were right, but that this was a time when being scientifically right was not really the point. Given the horror, physical and moral, that people had of venereal diseases, Kellogg was going to win any struggles going.

"Nature's Pruning Hook"

That war is dysgenic is the major worry of Anglo-American writing about the biological effects of warfare. Cambridge physicist-agriculturalist W. C. D. Whetham, who at the beginning of the war had had personal experience of recommending young men for junior officer roles, was appalled at the subsequent carnage among precisely this group of soldiers. The "brightest and best" were being wiped out. By the second year of the war, "a miserable remnant of the University alone was left in residence, made up of the war-useless old, the unsound in body rejected by the doctors, and the unsound in spirit, who, for one reason or another, had no stomach for the fight—a true survival of the worse" (Whetham 1917, 217). Of course, we need the generals to run the war, but eugenically they are a bit of a sideshow—"every promising subaltern [officer below the rank of captain] killed may racially be worth many Generals."

Another who spoke to this theme, in the early years of the war, was J. Arthur Thompson, Scottish holistic biologist and popular science writer. One can only be grateful for the morally uplifting effects of war. However: "Let us not seek to conceal the fact that war, biologically regarded, means wastage and a reversal of eugenic or rational selection, since it prunes off a disproportionately large number of those whom the race can least afford to lose" (Thompson 1915, 5). Although he used the same metaphors, the totally reverse philosophy was apparently held by another Scottish biologist, Sir Arthur Keith, anatomist and anthropologist and probably today best known for his possible involvement in the Piltdown Hoax, the discovery in 1912 in the South of England of a supposed "missing link" between humans and apes. Expressing sentiments he was already endorsing during the war, some years later he wrote: "Nature keeps her human orchard healthy by pruning; war is her pruning-hook" (Keith 1931, 50). Actually, taken out of context, this is as misleading as it is notorious. Keith never denied the bad effects of war considered from the perspective of the individual. Impressed—some might say overly impressed—by the success of the unified Germany through and after the Franco-Prussian War of 1870, not to mention a rather creepy conviction about the superiority of all things Scottish, Keith thought the key to evolution is the group rather than the individual. War plays a crucial role in this process, as he explained in 1915 to an audience at one of the London medical schools. "The title I gave to my lecture at St. Thomas's was 'War as a Factor in Racial Evolution.' I had traced the rise of local groups into tribes, and tribes into nations. The animosity that kept nations apart was the same evolutionary spirit which kept primitive groups apart. I illustrated my theme by the recent history of Germany—how she had used war to bind her constituent States together and to isolate herself from surrounding peoples" (Keith 1950, 396–397).

Endorsing some rather frightening social views—"Race prejudice, I believe, works for the ultimate good of mankind and must be given a recognized place in all our efforts to obtain natural justice for the world" (1931, 49)—after the Second World War, Keith had to do some rather fancy footwork on the subject of the Jews. Apparently, Hitler was right in spotting that the Jews are separate but wrong to think this a bad thing. "The Jewish family is now, and always has been, distinguished by the strength, warmth, and intensity of the bonds of affection and of love which unite all of its members into a cooperative whole" (1947, 65). Yet, Keith never changed his essential message. Humans are biologically a mixture of aggressive traits and altruistic traits, the former directed to those outside the group and the latter to those within the group. Keith was taken with the Mongolians. You don't get a fiercer warrior than a man from one of those tribes—"harsh, stern, ruthless, and vindictive towards his enemies." Simultaneously, a regular pussycat at home—"he loved, and was loved by, his family circle; he helped, entertained, and cherished his friends" (Keith 1947, 176). What then of the costs of war? "War is dysgenic." That cannot be denied. However, the costs to the individual are outweighed by the benefits to the group.

> How can war, which deprives a nation of its flower of manhood, serve any salutary or evolutionary purpose whatsoever? Let us suppose for a moment that our young men and women had shared the pacific ideals which had been preached to them by so many of our men of letters and had refused to bear arms, and so left England open to the enemy. England would then have ceased to be the home of a free and independent nation; the English as an evolutionary unit would have been erased from the map of Europe; their destiny would have passed into the hands of the enemy into the hands of Germany for a time at least. In war, then, a nation sacrifices its best blood in order that it may maintain its integrity and perpetuity, and so fulfills its destiny under the law of evolution. (145)

Keith shared the bishops' sentiment that it is a noble and fitting thing that the young of the nation be allowed to make the supreme sacrifice for us all. Looking ahead to the Second World War, by February 1944 nearly forty thousand young men (average age twenty-two) had died in bombing raids over Germany. (The final count was nearly sixty thousand.) Too bad, but let us be realistic. "The youth of a nation sacrifices itself in war in order that its nation may live, move, and have its being, remaining safe and free to work out its own evolutionary destiny. Youth pays the premium of national safety and integrity. That truth, although often forgotten, is freely acknowledged by those who have studied the evolutionary effects of war" (211). God and Darwin in harmony on this crucial point.

Scottish Common Sense

The thinking of Friedrich von Bernhardi loomed over any evolution-based British discussion of the Great War. Locked in an apparently mortal conflict with Germany, von Bernhardi's ferocious reading of Darwin seemed all too real, or at least pertinent for understanding what was driving German aggression. Return to Peter Chalmers Mitchell. He is today a forgotten figure. He was not one of the leading evolutionists, nor would he have made such a claim on his own behalf. The third of our Scottish biologists, he was an administrator, secretary of the Zoological Society of London. A highlight of his very successful thirty-year rule was the conception and organization of an out-of-town, park zoo, Whipsnade. Chalmers Mitchell was a humane man with a broad interest in culture, and because this extended so deeply into German culture, it was nigh foreordained that he would take on von Bernhardi, full frontal as it were. *Evolution and the War* (1915) pulls no punches.

Chalmers Mitchell's first move was to point out that when Darwin spoke of a "struggle" for existence, as often as not this was metaphorical and had little or nothing to do with actual physical struggle—naked mud wrestling as it were. It was often much more a matter of finding and exploiting niches that were unknown to or unusable by others. Cockroaches make this point nicely. At the zoo, you have three kinds: Oriental, German, and American. They have lived there happily for years. "The large American cockroach infests the Reptile-house, the German species the Small Bird-house and the Ape-house, and the oriental insect is specially addicted to the Small Mammals'-house and the keepers' rooms. I can find no evidence that the members of the different species attack one another . . ." (Chalmers Mitchell 1915, 33). This led to the more general conclusion: "Looking through the animal kingdom as a whole, and remembering that the vegetable kingdom is as much subject and responsive to whatsoever may be the law of organic evolution, I find no grounds for interpreting Darwin's 'metaphorical phrase,' the struggle for existence, in any sense that would make it a justification for war between nations" (41).

Chalmers Mitchell turned now to our species, tracing its evolution and speaking of the racial traits we find separating peoples. His point was simply that biology and culture do not map one on one, and hence there is no reason to think that nations have been formed by or mainly by biological processes, like the struggle for existence. Of course nations are entities, but—he clearly had Spencer in his sights, although Keith would also be under fire here—analogies with organic individuals are thin to nonexistent. "Even if the struggle for existence were the sole law that had shaped and trimmed the tree of life, it does not necessarily apply to the political communities of men, for these cohere not because of common descent but because of bonds that are peculiar to the human race" (64). In other words, nations are formed and maintained by politics and ideas and so forth, rather than by physical strife. Talking of the present situation: "In this sense the struggle between the nations is in truth a war of culture, a resistance by England and France, Russia and Belgium to the attempt to force on the world one particular conception of civilization." And he didn't buy into the idea that because German is best that means all should imitate or be forced to copy. "May I say in passing that even if we were to accept the German view that German 'Kultur' leads to the highest ideal of civilization, submission to it would be no less a crime against the human race. We require variety, different ideals among which to choose, and freedom to make our choice" (67).

Interestingly, the thoroughly commonsensical Chalmers Mitchell—he took great pride in being a Scot, although not in the tribal fashion of Keith—had as little patience with the eugenicists and their sepulchral warnings as with the German war makers. He wasn't at all sure that the killing in war meant the inevitable decline in a nation's fortunes. "Those who survived the perils and privations of service were presumably in many cases the most active and rugged; the weaker portion having succumbed in the meanwhile either to wounds or sickness. The result was that the generation conceived directly after the war was as much above the average, especially evinced in general physique more than in stature, as their predecessors, born of war-times, were below the normal" (77). Continuing this theme and almost cheekily making a personal stab at the von Bernhardis of the world. "No doubt it may be urged that many of the bravest are likely to perish in war; but many of those whose natural disposition is martial must also disappear." In other words, there is a clearing out of the hotheads. Indeed, "the inference might be drawn that forty years of unbroken peace have made Germany too ready to embark on war, as there has been no opportunity for her fire-eaters to be eliminated" (78).

Chalmers Mitchell had a larger agenda that covered both German militarists and eugenicists, namely to take humankind out of the strictly biological and to argue that now we have escaped up into culture where we can make decisions about ourselves. One senses something very Kantian (a philosopher for whom Chalmers Mitchell showed considerable understanding and sympathy) about all of this, with humans no longer puppets of our past but in some sense autonomous beings in charge of our own destinies. "Writing as a hard-shell Darwinian evolutionist, a lover of the scalpel and microscope, and of patient, empirical observation, as one who dislikes all forms of supernaturalism, and who does not shrink from the implications even of the phrase that thought is a secretion of the brain as bile is a secretion of the liver, I assert as a biological fact that the moral law is as real and as external to man as the starry vault" (107). Note the key causal factor. "Here it is culture that plays the vital role. It is the work of the blood and tears of long generations of men. It is not in man, inborn or innate, but is enshrined in his traditions, in his customs, in his literature and his religion." Mitchell is as pro-progress as the rest of them, but for the right reasons. Culture's "creation and sustenance are the crowning glory of man, and his consciousness of it puts him in a high place above the animal world. Men live and die; nations rise and fall, but the struggle of individual lives and of individual nations must be measured not by their immediate needs, but as they tend to the debasement or perfection of man's great achievement." Chalmers Mitchell saw with clear eyes what the Germans were up to and he just didn't want his hero Charles Darwin tainted by any of it.

American Reaction

In America too there was opposition to Germanic use of Darwinian themes. Jacques Loeb, groundbreaking physiologist, German-born but by the time of the war a long-time American professor, showed little enthusiasm for Darwinism generally and even less for extensions to human societies. "The terms 'survival of the fittest' or 'struggle for existence' were never more than poor metaphors to express the fact that the chemical compounds required for the growth of organisms are restricted in quantity and that as a consequence unlimited reproduction of organisms is impossible" (Loeb 1917, 75). They simply don't apply to human conflicts. "The methods by which the stronger 'conquer' weaker nations have nothing in common with the fact that salt water fish die when put into fresh water or that microorganisms can not multiply unless they have their proper culture medium." In any case, humans aren't the sorts of beings to which these sorts of metaphors inevitably apply. "Of course, there are animals which are as brutal and predatory as the war enthusiasts think human beings should be—but this is a different thing from calling this brutality a universal law of living nature. Fortunately the normal human being does not belong to this brutal type." Loeb managed also to put in a nasty dig (without naming names) at the eugenicists like Kellogg who feared the deleterious biological effects of war. "Compared with the misery and anguish, the general loss of life and of liberty, and the economic waste caused by war, the possible hereditary effects on the population, if there are any, are too trivial to be mentioned" (73).

Meanwhile, Vernon Kellogg himself was evolving. Whatever one may think of his thinking about evolution and its causes, he was a man of enviable moral stature. Remember that, as soon as the war broke out, Kellogg moved to Belgium to work in the cause of humanitarian relief. This lasted until America itself entered the war in 1917. By this time, Kellogg had changed. During his time in Belgium, he quartered with the German officers, meeting the head of the German chief of staff Erich von Falkenhayn and even the "All-Highest" (Kaiser) himself. Long into the night were the conversations about the war and about the causes and the philosophies which drove men forward. Kellogg reported on ideas being promoted. "The creed of the Allmacht of a natural selection based on violent and fatal competitive struggle is the gospel of the German intellectuals; all else is illusion and anathema" (Kellogg 1917, 28). As always, there are value components, with the military lauding the supposed progressive benefits, however cruel the process. "This struggle not only must go on, for that is the natural law, but it should go on, so that this natural law may work out in its cruel, inevitable way the salvation of the human species. By its salvation is meant its desirable natural evolution. That human group which is in the most advanced evolutionary stage as regards internal organization and form of social relationship is best, and should, for the sake of the species, be preserved at the expense of the less advanced, the less effective" (29). No prizes for guessing which human group (in the view of the High Command) has achieved "the most advanced evolutionary stage."

The result of listening to this sort of stuff night after night was a little book, *Headquarters Nights*, within which Kellogg wrote of his move from pacifism to the conviction that Germany must be fought and changed. There is need of "fighting this war to a definitive end—that end to be Germany's conversion to be a good Germany, or not much of any Germany at all. My 'Headquarters Nights' are the confessions of a converted pacifist" (23). Like Peter Chalmers Mitchell, Kellogg agreed that humans have evolved naturally, but (also like Chalmers Mitchell) Kellogg insisted that the struggle for existence (which, for all of his doubts about Darwinism, Kellogg was prepared to accept at some level) is not necessarily bloody in the sense presupposed by the Germans.

Although there was no specific reference to Kropotkin (nor, to be fair, was there any compelling reason why there should be), Kellogg made exactly his point about the struggle being often with the environment and not with other organisms. "For example, the voluntary or involuntary migration of representatives of a species hard pressed to exist in its native habitat, may release it from the too severe rigors of a destructive climate, or take it beyond the habitat of its most dangerous enemies, or give it the needed space and food for the support of a numerous progeny" (26).

Truly, Kellogg was not interested in making a biological case against the Germans. Revealingly, all references to the dysgenic effects of warfare were dropped. He was interested in making a moral case. It was the devastation of Belgium and Northern France that appalled him. He simply could not believe the horrors inflicted on these wretched folk by the Germans. Kellogg stressed that he hated war even more than when he was a pacifist, but the line had to be drawn somewhere and sometime, and that somewhere and sometime was here and now. Writing of his travels through the blighted lands: "The sights and the incidents of those trips are too harrowing to exploit. They are untellable, intimate memories for us, but they went far in making us convinced and bitter believers that the only comprehensible answer to the German philosophy of 'raison d'Etat,' and 'military exigency,' to these ravages of non-combatant countryside and village, is an answer of force" (52–53). It was these deaths of the civilians and the deportations of Belgians to Germany, where they were forced to work for the Fatherland, that most upset Kellogg. Acts like these are against any code of war and fill one with revulsion. Kellogg recognized human traits in some of the Germans, but overall they are frightening and their culture is even worse.

Kellogg like Mitchell knew Germany from firsthand experience—remember, he had studied there and hence his ability to speak the language, something vital to his war work. Yet, one does not sense the appreciation of aspects of German culture one finds in the writing of the Scot. "Even were German Kultur that most desirable thing that the German Intellectuals have said it is—and that most of us are convinced it is not—the Germans are utterly unable to make it over to any other people." With a dreadful premonition of what was to come, Kellogg added: "German Kultur stifles the good in man for the good of a man-made Juggernaut called the State" (80–81).

Aftermath

On the morning of November 11, 1918, the dreadful war came to a halt. Hardly an end, for it was an armistice and much was unresolved. Germany was beaten but bitter, and the peace negotiations, ending with the Versailles Treaty, simply exacerbated things (MacMillan 2002). The Treaty was blamed—with good reason if not as much good reason as many have thought—for many things that were to happen in the next twenty years, ending with the start of the Second World War, something in major respects the continuation of the First World War. The succession of events ended finally only in 1945, with the unconditional surrender of the Nazis' Third Reich (Kershaw 2015).

We see the incomplete nature of the ending of the First World War in the writings of those involved in the debate during the war. Von Bernhardi was, as they say, bloody but totally unbowed. The loss of the war by Germany was entirely a function of mismanagement, and the Versailles Treaty limiting Germany's right to a standing army is nothing short of disgraceful. Bernhardi's philosophy of strife was unchanged. "Always, as long as humans are humans, force in its most encompassing sense will indicate the political and cultural meaning of states. It is at the root of all mental and moral progress." Note the emphasis on progress. War does not just lead to change. It leads to change of a good kind, namely upward. In such aspiration, however, von Bernhardi could not resist a stab at the present failures. "I hope that the German man who now seems sunken in selfishness and self-indulgence will improve and time will find a purified people which will prove worthy of its ancestors and also consider war, like reality has designed it" (von Bernhardi 1920, 237, quoted in translation by Krischel 2010, 147).

Kellogg's change during the war was lasting. His short, immediate post-war book, *Germany in the War and After* (1919), is mainly about the sad state of Germany, from about 1916 to the present—lack of foodstuffs, machinery wrecked, trains not on time. It is informative and remarkably non-emotive. Kellogg makes clear his distaste for the German political culture that he believed led to the war, but his stance is not one of vindictive crowing. More an assessment of the present and the needs that must be filled. However, eugenics is gone completely and explicitly. This is a man who has turned from Darwinism in society in any form. "My own experience in the last four years has done much to make me over from a convinced believer in the dominating influence of heredity over environment and education in determining human behavior and moral makeup into a believer in the great possibilities of the modifying effect of environmental conditions" (Kellogg 1919, 95–96). It is culture over biology all of the way. "Germans are not so different from Englishmen and Americans by stock, that is, by inheritance; but they have been made different by education, using the word in its larger biological sense. And if this is true there is hope that they may be changed again if their education is sufficiently changed; that if their environment becomes strongly democratic and democratizing they may in time become of the democratic faith" (96). In other words, with the right education and a modicum of goodwill and luck, we might succeed in turning Germans into Americans. "Can Germans come to this? Not if human nature is immutable. But it is not. The present human nature of the Germans is not necessarily a human nature they have always had. Indeed it is certain that it is not" (97).

In terms of the perceived relationship between today's humans and our ancestors, the hominins (in our language), Kellogg was firmly on the Darwinian side. Evolution including selection (by whatever name) has led up to humankind. Now, thanks to culture, we have escaped our biology. We have changed. "But the important element which has made possible the immense hastening of this change of nature, of mental and moral makeup, has been the element of environment and education, rather than that of pure natural selection" (98). It is the cultural that counts. "In our social evolution we have been able to hold fast, by virtue of speech and writing, to steps which are not actually a part of our natural evolution. We have a social or traditional inheritance as well as a physical inheritance. And it is by conscious modification of our environment and education that we can determine the character of this all-important influence on our lives." Germany has great potential and it can change. "All they need is the proper education; the kind of environment that the world has come to understand as the best for right influence on human evolution" (100–101).

We move on now to the period from the end of the First War to the end of the Second War. As we do, see how our themes continue to play out. The Christians, with the obvious exception of the pacifists, who were very much a minority, saw war as a result of our sinful nature. It exists and will surely go on existing. There are times when it is our Christian duty to fight and to combat evil. We are always in the hands of Providence. Whatever we do cannot be the last word, and whatever we do must be put in the context of God's immanent power and concern for His creation. Almost paradoxically, given that both sides start from the shared presumption about our thoroughly unpleasant, violent nature, the Darwinians saw things in an entirely reverse way. They hated war now, but saw it as part of the natural process leading progressively up to the evolution of humankind. Hence, they could not at all condemn war universally. However, they were looking for its end, thinking it to be an appalling thing—for many now it was counter-evolutionary and leading away from the good. They looked both biologically and culturally for factors spelling this end. This applies even to someone like Kellogg who changed his mind so drastically. He wanted war to end, and thought that culture makes this possible, even though he thought also that this war might be necessary. The exception of course are the German militarists, using Darwin for their own conflict purposes. But we have seen that German thought was never truly soaked in Darwinian ideas and always owed as much if not more to its own philosophies of change. So even here we don't really have an exception to the paradox.

7 | Realists and Pacifists

THE FIRST WORLD WAR ENDED because Germany was exhausted. It could fight no longer. The wounds continued to fester, and, as the 1930s gathered steam, with the coming of fascism to Italy in the decade before; with the horrendous civil war that engulfed Spain; and above all with the rise of Hitler and his Third Reich, many started to fear that another world war was on the horizon and that we must prepare for this. It was a message heard increasingly by many Christians, who found their spokesperson in the already-introduced American Lutheran minister, Reinhold Niebuhr. He went from pacifism to a position that was labeled "Christian realism," meaning one can and must combine Christian commitment with a realistic appraisal of the nature of the world, and realization that sometimes the only right course of action leads to war.

Christian Realism

Niebuhr's classic *Moral Man and Immoral Society* lays out the central theme in the opening words. "The thesis to be elaborated in these pages is that a sharp distinction must be drawn between the moral and social behavior of individuals and of social groups, national, racial, and economic; and that this distinction justifies and necessitates political policies which a purely individualistic ethic must always find embarrassing" (Niebuhr 1932, 139). He goes on to say: "Individual men may be moral in the sense that they are able to consider interests other than their own in determining problems of conduct, and are capable, on occasion, of preferring the advantages of others to their own." Unfortunately, when it comes to the group, to society, things are otherwise. Moral "achievements are more difficult, if not impossible, for human societies and social groups. In every human group there is less reason to guide and to check impulse, less capacity for self-transcendence, less ability to comprehend the needs of others and therefore more unrestrained egoism than the individuals, who compose the group, reveal in their personal relationships."

One can see already how Niebuhr makes the case for war. Individually, a Christian may try to follow the Sermon on the Mount. Collectively, and hence as a citizen, this may not be possible and fighting as a soldier may be necessary. May even be obligatory. None of this really should come as a surprise. Niebuhr is a Lutheran theologian. Apart from the fact that he gives the state authority that not everyone thinks entirely wise, that means he is deeply committed to original sin, to the notion of Providence, and—as we have seen already—to an almost visceral dislike of ideas of progress. That truly is to follow a false God. "Human nature is limited and hence Progress is impossible. Whatever increase in social intelligence and moral goodwill may be achieved in human history, may serve to mitigate the brutalities of social conflict, but they cannot abolish the conflict itself. . . ." Niebuhr has little time for "educators who emphasise the pliability of human nature, social and psychological scientists who dream of 'socialising' man and religious idealists who strive to increase the sense of moral responsibility" (148). Condescendingly, he allowed that they "can serve a very useful function in society in humanizing individuals within an established social system and in purging the relations of individuals of as much egoism as possible." But let us be realistic. "In dealing with the problems and necessities of radical social change they are almost invariably confusing in their counsels because they are not conscious of the limitations in human nature which finally frustrate their efforts" (Niebuhr 1932, 148).

It is Niebuhr's contemporary, the ardent Darwinian Pragmatist, John Dewey, who is the target here. He and his fellow educational theorists make false or unrealizable promises of a harmonious better tomorrow. Their ideas are doomed. "While this hope of the educators, which in America finds its most telling presentation in the educational philosophy of Professor John Dewey, has some justification, political redemption through education is not as easily achieved as the educators assume" (302–303). As the unsatisfactory results of the last war show only too clearly, nothing significant will ever change. "Thus society is in a perpetual state of war. Lacking moral and rational resources to organise its life, without resort to coercion, except in the most immediate and intimate social groups, men remain the victims of the individuals, classes and nations by whose force a momentary coerced unity is achieved, and further conflicts are as certainly created" (164). And don't think that milk-sop religion is going to help all that much. "The demand of religious moralists that nations subject themselves to 'the law of Christ' is an unrealistic demand, and the hope that they will do so is a sentimental one" (204). The individual-group tension comes right into play here. "Even a nation composed of individuals who possessed the highest degree of religious goodwill would be less than loving in its relation to other nations. It would fail, if for no other reason, because the individuals could not possibly think themselves into the position of the individuals of another nation in a degree sufficient to ensure pure benevolence."

Niebuhr agreed that God in His Providential role will make all right. This should not be understood in a day-to-day pragmatic sense. It is an eschatological doctrine, about end times. "The cross is the symbol of love triumphant in its own integrity, but not triumphant in the world and society. Society, in fact, conspired the cross." War, therefore, is no surprise. "To the end of history the peace of the world, as Augustine observed, must be gained by strife" (Niebuhr 1932, 335). That means that, as a general philosophy, Niebuhr has little time for pacifism. Pacifists advocate nonresistance, thinking this will make for a better world. "Their conviction is an illusion, because there are definite limits of moral goodwill and social intelligence beyond which even the most vital religion and the most astute educational programme will not carry a social group, whatever may be possible for individuals in an intimate society" (164).

Actually, Niebuhr had much respect for traditional peace churches, like the Mennonites. He thought they did serve a purpose in showing us an ideal. Generally, as he argued in an essay that appeared when the Second World War was started (but before America joined in), pacifism is impossible and dangerous and basically immoral (Niebuhr 1940). "Most modern forms of Christian pacifism are heretical. Presumably inspired by the Christian gospel, they have really absorbed the Renaissance faith in the goodness of man, have rejected the Christian doctrine of original sin as an outmoded bit of pessimism, have reinterpreted the Cross so that it is made to stand for the absurd idea that perfect love is guaranteed by a simple victory over the world" (Niebuhr 1940, 239). Pacifism of this kind is Christian heresy and profoundly simplistic when seen through the lens of human nature judged in a purely secular fashion. Again, Providence is invoked and again it is argued that this is not something in human time. "The New Testament does not . . . envisage a simple triumph of good over evil in history. It sees human history involved in the contradictions of sin to the end. That is why it sees no simple resolution of the problem of history." Of course, the Kingdom of God will bring about an eventual resolution of the human dilemma. Just not as we might expect it. "The grace of God for man and the Kingdom of God for history are both divine realities and not human possibilities."

Niebuhr stands in the Augustinian-Lutheran tradition, but it is important to make clear that his is not truly a just war position. A perceptive critic—Keith Pavlischek, a military affairs expert—points out that Niebuhr, in rejecting his early pacifism, thought he was moving on beyond the theology of his early days. He was leaving a theology of the Social Gospel that puts heavy emphasis on the Jesus of the gospels, of the Sermon on the Mount. Yet, in many key respects, Niebuhr continued in this tradition. "Niebuhr and his disciples conceded that the Gospel ethic, or the ethic of Jesus, is a pure ethic of love and nonviolence, but concluded that the perfect morality modeled by Jesus (and expected of his disciples) is not practical in human society and thus must be moderated by a pragmatic or realistic ethic of responsibility that requires a choice of lesser or necessary evils on behalf of the community" (Pavlischek 2008, 55).

Here's the main point of difference. For Niebuhr, when you go to war, in a sense you are caught in evil. Your duty as a citizen clashes with your obligations as a Christian individual. This is not just war theory, which explicitly argues that in fighting in a just war you are acting fully as a Christian individual. The critic continues: "The classical just war tradition shares with the pacifist the belief that if Jesus condemns an action as vicious and otherwise prohibits his disciples from engaging in that action, or if Holy Scripture more generally prohibits an action, then Christians are obligated to refrain from that action." This is why Augustine goes to so much trouble to show that the Centurion implies that one can, as a Christian, go to war. "Niebuhr's exegetical strategy repudiates the interpretation of both classical Christian just war teaching and historic Christian pacifism. Classical Christian just war theologians all explicitly reject the notion that the Sermon on the Mount embodied an 'ideal ethic' for the Christian, insisting that Jesus intended to restrain personal or private vengeance, not to restrain the just employment of political force, which to the contrary, is understood as an agent of his Father's wrath and love" (57–58). Romans 13 is the definitive text. "1. Let every soul be subject unto the higher powers. For there is no power but of God: the powers that be are ordained of God." Continuing: "7. Render therefore to all their dues: tribute to whom tribute is due; custom to whom custom; fear to whom fear; honour to whom honour." If you are told to fight, then as a good Christian, you fight.

The point is simple. All agree that, at the individual level, Jesus expects us to be loving and nonviolent. Niebuhr's "realist" position is that when we get into groups, we have to go against this. In some sense, war is necessary but sinful. The just war tradition is that, in groups, Jesus lifts the prohibition against violence. War in itself is a bad thing and comes about because of original sin. But there are times when, morally and religiously, in the name of Jesus, we ought to go to war. In the words of the critic, "early Christian approaches to war treated it as a necessary evil. Each held that the person who used just force was acting in a way consonant with God's wishes and was, though in a way less praiseworthy than bishops and clerics, following Christ. The just soldier's acts in war were thus thought to be positively good acts—acts that would shape him into the kind of person fit for beatitude with God" (58). Adding: "For Saint Augustine, the decision to go to war is never a decision between two evils. Hence, in Contra Faustum 22.74, Saint Augustine famously asks, 'What is the evil in war?' "

Karl Barth

Many would regard Reinhold Niebuhr as the most important American theologian—certainly the most important, American, Protestant theologian—of the twentieth century. A major influence on his thinking, especially in his moves in the 1920s from the more liberal theology with which he grew and developed, were the writings of he whom many would describe as the most important European theologian—perhaps simply the most important theologian—of the twentieth century, the Swiss pastor and then professor, Karl Barth. As it happens, perhaps not entirely surprisingly in one who was trying to make his own distinctive way, Niebuhr tended to be a bit contemptuous about Barth. Supposedly, the Swiss thinker refused to engage politically in the world, and "Karl Barth actually reduced Lutheran pessimism to a new level of consistency and made it even more difficult for the Christian conscience to express itself in making the relative decisions which are so necessary for the elaboration of justice in the intricacies of politics and economics." Adding, a little ungraciously: "Karl Barth's present position of uncompromising hostility to Nazism cannot change the fact that his system of thought helped at an earlier date to vitiate the forces which contended against the rising Nazi tyranny" (Niebuhr 1940, 58). At the end of the Second World War, after people learned of Dietrich Bonhöffer's death, Niebuhr paid an appropriately moving tribute.

Yet, he could not resist the jab that when Bonhöffer was a student in America in 1930–1931, "he was inclined to regard political questions as completely irrelevant to the life of faith. But as the Nazi evil rose he became more and more its uncompromising foe. With Barth he based his opposition to Nazism upon religious grounds" (Niebuhr 2015b, 656).

Apart from the irony of criticizing theologians because their position is one of religion and theology, this is surely unfair to Barth and his followers. Not only is there the example of Bonhöffer, Barth himself in 1934 wrote the Barmen Declaration that, in the name of Christ, stood firmly against the Nazis. "Through him befalls us a joyful deliverance from the godless fetters of this world for a free, grateful service to his creatures. We reject the false doctrine, as though there were areas of our life in which we would not belong to Jesus Christ, but to other lords—areas in which we would not need justification and sanctification through him" (Barth 1934, 8.14). Moreover, whether direct influence or not, we see major elements of Niebuhr's thinking there in the earlier writings of Barth. During the First War, Barth, then a Swiss pastor, was appalled at the way in which his German teachers, Adolf Harnack in particular, enthusiastically became cheerleaders for the German cause. Especially troubling was their signing the Manifesto of the Ninety-Three, defending Germany against charges of misconduct in the entry into and occupation of Belgium. Barth ripped up his theology, as it were, and started again—in the later years of the war, writing a detailed commentary on Paul's Epistle to the Romans.

Barth insisted that Christians return to their roots, as based on readings of the Bible. By the end of the nineteenth century, liberal theologians—particularly those pushing the earlier-mentioned "Social Gospel"—had made not just Jesus but even God one of the chaps. He was a kind of benevolent Big Brother, in there helping you to improve society. Not so much of the Providence, and much more of the progress—something of course going on in Anglican circles also. Barth would have none of it. God is unknown, wholly other. "What is clearly seen to be indisputable reality is the invisibility of God, which is precisely and in strict agreement with the gospel of resurrection—His everlasting power and divinity. And what does this mean but that we can know nothing of God, that we are not God, and that the Lord is to be feared?" (Barth 1933, 46–47) There is an (acknowledged) debt to Kierkegaard here. We do not reason to God—Barth had no time for natural theology. We accept on faith, even though—precisely because—it might be absurd. Faith confirmed by reason is empty. It requires a risk. "The Gospel does not expound or recommend itself. It does not negotiate or plead, threaten, or make promises. It withdraws itself always when it is not listened to for its own sake. 'Faith directs itself towards the things that are invisible. Indeed, only when that which is believed in is hidden, can it provide an opportunity for faith.' " Reason has no role—can have no role. "The gospel of salvation can only be believed in; it is a matter for faith only" (39). It all starts to get very Augustinian. "If thy life be without that justification which God alone can give, it is utterly devoid of any justification at all" (73). Without God's grace, we are lost. "For there is no distinction: for all have sinned, and fall short of the glory of God; being justified freely through his grace by the redemption that is in Christ Jesus" (Romans, 3, 22b–24). Thoughts of progress are chimerical. "Are our affirmations and negations anything more than chance or whim? When we judge one man a criminal and another a saint; when we destine one to hell and another to heaven; when we believe 'good will grow better and better, and evil worse and worse' (Harnack), are we not purely capricious?" (226).

What then of Barth's actual position on war and its morality? Not surprisingly, given that it was the war-mongering theologians who sparked his taking a very different path from the theology of his day, there is a deep empathy toward pacifism. In Barth's most detailed discussion of war—from the *Church Dogmatics*, published in 1951—he makes crystal clear the horror of armed conflict. There used to be a certain nobility in conflict. "Today, however, the increasing scientific objectivity of military killing, the development, appalling effectiveness and dreadful nature of the methods, instruments and machines employed, and the extension of the conflict to the civilian population, have made it quite clear that war does in fact mean no more and no less than killing, with neither glory, dignity, or chivalry, with neither restraint nor consideration in any respect" (Barth 1951, 453). This said, there is a place for war in the Christian world. Sometimes it is necessary to take up arms. Specifically, while Barth had little time for the First War, he thought fighting the Second World War was necessary for the allies. He wrote an open letter to British Christians in 1941: "Even though we may lose, we must fight to the bitter end the evil that Adolf Hitler and his Third Reich represent" (Barth 1941, n.p.) In an important passage, he continues: "it is the clear will of God that we should recognize the nature and power of the movement, in order to combat it with all of our strength. The obedience of the Christian to the clear will of God compels him to support this war."

Why important? Because it shows the reasoning behind Barth's thinking. He rejects just war theory. He thinks that it is part of the natural theology that he as a Christian must reject. "All arguments based on Natural Law are Janus-headed. They do not lead to the light of clear decisions, but to the misty twilight in which all cats become grey" (Barth 1941, n.p.). He wants nothing to do with the "liberty of the individual" and "social justice" and the like. "I am disturbed by the fact that these conceptions are concerned with principles which might also be those of a pious Hindu, Buddhist or Atheist: and that, however beautiful and fruitful they may be, they do not touch at all on the peculiarly Christian truths on which the Church is founded." The Resurrection of Christ! It is to this that we turn for the grounding of our beliefs about war—"the ultimate reason I put forward for the necessity of resisting Hitler was simply the resurrection of Jesus Christ." In other words, that in fighting Hitler we are doing the right thing, we have faith and assurance directly from God. No arguments are needed once we accept the sacrifice on the Cross and the subsequent conquering of death. Barth stands firmly against his Christian mentors who he feels did so badly in the First War; but, although he does not agree with Niebuhr's case for the necessity of war, he agrees with Niebuhr in agreeing that the Christian can and might (on religious grounds) have to wage war, and also he agrees with Niebuhr in thinking that these grounds are not just war theory.

Pacifism

> I will myself do the best I can to settle my account with the Unknown Soldier. I renounce war. I renounce war because of what it does to our own men. I have watched them come in gassed from the front-line trenches. I have seen the long, long hospital trains filled with their mutilated bodies. I have heard the cries of the crazed and the prayers of those who wanted to die and could not, and I remember the maimed and ruined men for whom the war is not yet over. I renounce war because of what it compels us to do to our enemies, bombing their mothers and villages, starving their children by blockades, laughing over our coffee cups about every damnable thing we have been able to do to them. I renounce war for its consequences, for the lies it lives on and propagates, for the undying hatreds it arouses, for the dictatorships it puts in the place of democracy, for the starvation that stalks after it. I renounce war and never again, directly or indirectly, will I sanction or support another! 0 Unknown Soldier, in penitent reparation I make you that pledge. (Fosdick 1934, 97–98)

This is from a sermon to the Unknown Soldier, given on November 12, 1933, by Harry Emerson Fosdick. He had not always been a pacifist. He supported the American entry into the First World War and went to France to be with and support the troops. A little volume, *The Challenge of the Present Crisis* (1917), ended with a quotation from a French mother to her son in Canada. "My dear boy: You will be grieved to learn that your two brothers have been killed. Their country needed them and they gave everything they had to save her. Your country needs you, and while I am not going to suggest that you return to fight for France, if you do not return at once, never come" (Fosdick 1917, 98–99). To which, Fosdick added: "Multitudes are living in that spirit today. He must have a callous soul who can pass through times like these and not hear a voice, whose call a man must answer, or else lose his soul. Your country needs *you*. The Kingdom of God on earth needs *you*. The Cause of Christ is hard bestead and righteousness is having a heavy battle in the earth—they need *you*" (99).

Then, the war over and looking at the devastation of the land and the mowing down of the young, Fosdick had a complete change of mind. By 1925, he was preaching about the idea of going to war to protect the weak. Rhetorically and ironically, he noted: "See how modern war protects the weak: 10,000,000 known dead soldiers; 3,000,000 presumed dead soldiers; 13,000,000 dead civilians; 20,000,000 wounded; 3,000,000 prisoners; 9,000,000 war orphans; 5,000,000 war widows; 10,000,000 refugees. What can we mean—modern war protecting the weak?" (Fosdick 1925, 7) There is no direct, explicit appeal to just war theory, but Fosdick is grounding his argument on one of its essential premises, that war can never be justified if the end is worse than the beginning. Fosdick's arguments, however, were never philosophical. They were always those of a Christian minister. "We cannot reconcile Jesus Christ and war—that is the essence of the matter. That is the challenge which today should stir the conscience of Christendom." War, he continued, is the antithesis of the Christian message. It is "the most colossal and ruinous social sin that afflicts mankind; it is utterly and irremediably unchristian; in its total method and effect it means everything that Jesus did not mean and it means nothing that He did mean; it is a more blatant denial of every Christian doctrine about God and man than all the theoretical atheists on earth ever could devise" (18). Fosdick never wavered from this. He was a pastor to his flock. In the Second World War, his church went out of its way to support and show love toward its young men who went out to fight. This did not change his own mind. His pacifism was absolute and strong.

An Englishman much inspired by Fosdick was the Reverend Hugh Richard Lawrie "Dick" Sheppard, Anglican priest and for a while Dean of Canterbury. He too served as a pastor in France, in the First World War, and he too turned to pacifism. Triggered by Fosdick's Unknown Soldier sermon, Sheppard and others formed the (British) Peace Pledge Union (PPU). A nonsectarian organization, it is open to all who can sign its pledge. "I renounce war, and am therefore determined not to support any kind of war. I am also determined to work for the removal of all causes of war." Sheppard wrote: "I was not a pacifist in the first year of War: as a professing Christian I ought to have been." Continuing: "War must be abolished if civilisation is to endure, and it is the duty of all honest and thoughtful men to resist war and the preparations and policies that make war imminent, even if in doing so they are accused of behaving 'unpatriotically.'" Like Fosdick, ultimately for Sheppard it was all a matter of the conscience of the Christian. "I maintain, with my whole soul, that the Church of Christ is not worthy to represent its Lord today unless it declares, without any equivocation or delay, that no leader or ranker under its banner may kill his fellow, his brother. Why? One answer will suffice: Christ would not permit it" (Sheppard 1935, 57).

Not all in the PPU shared this Christian commitment, and the organization numbered among others Bertrand Russell. Also members were the composer Benjamin Britten; the novelist Aldous Huxley; and two authors of the most-powerful accounts of the Great War, Siegfried Sassoon, author of the trilogy beginning with *Memoirs of a Fox-Hunting Man*, who fought bravely and then spoke publicly against the War, and Vera Brittain, who lost friends, brother, and fiancé in the conflict, and whose *Testament of Youth* is one of the greatest books of the twentieth century. In the late 1930s, with some reason, the PPU was regarded not just as a vehicle for appeasement but something so close to empathy with Hitler's aims that it was parallel to fascist organizations, with George Orwell accusing it of "moral collapse." More recently, and perhaps more to its credit, it incurred the wrath of Margaret Thatcher when it opposed her enthusiasm for military forays.

Fundamentalism

Expectedly, not all interwar pacifists thought alike. Fosdick was a liberal thinker, who was comfortable with science; and in particular was an enthusiast for evolution, although it seems clear that it was some kind of theistic form, where God has a hand in seeing that not all is blind and undirected. The evolution of humankind did not happen by chance. Expectedly, after the First World War, with the rise of Fundamentalism, Fosdick sailed into turbulent waters. William Jennings Bryan, the "Great Commoner," three times Democratic candidate for the presidency, was in 1913 appointed Secretary of State by then-president Woodrow Wilson. He resigned a couple of years later, when Wilson reacted harshly toward Germany at the sinking of the *Lusitania*, causing the deaths of 1,198 passengers and crew, of whom 128 were American citizens. After the War, with one of Bryan's targets, alcohol, now banned, thanks to a constitutional amendment enacting prohibition, he turned to his other main target, evolution. Bryan had read and been much moved by Vernon Kellogg's *Headquarters Nights*. In particular, he was appalled at the supposed Darwinian thinking of the German general staff. "The creed of the Allmacht of a natural selection based on violent and fatal competitive struggle is the gospel of the German intellectuals; all else is illusion and anathema." Continuing (and still speaking in the voice of his presumed German thinkers): "But as with the different ant species, struggle—bitter, ruthless struggle—is the rule among the different human groups. This struggle not only must go on, for that is the natural law, but it should go on, so that this natural law may work out in its cruel, inevitable way the salvation of the human species." Salvation is bound up with evolution, and that of its most "desirable" form. "That human group which is in the most advanced evolutionary stage as regards internal organization and form of social relationship is best, and should, for the sake of the species, be preserved at the expense of the less advanced, the less effective" (Bryan 1919, 303). The winners can and should impose their "Kultur" on the rest of us.

For the five years remaining in his life, Bryan lectured and wrote extensively against the vile science on which this kind of thinking is based. He found fault with Darwinism as science. "While 'survival of the fittest' may seem plausible when applied to individuals of the same species, it affords no explanation whatever, of the almost infinite number of creatures that have come under man's observation. To believe that natural selection, sexual selection or any other kind of selection can account for the countless differences we see about us requires more faith in chance than a Christian is required to have in God." Of course, it was never evolution as such that scared people like Bryan. If warthogs had had proto-warthogian ancestors, he could have lived with that. It was humankind that was at real issue. "Is it conceivable that the hawk and the hummingbird, the spider and the honey bee, the turkey gobbler and the mocking-bird, the butterfly and the eagle, the ostrich and the wren, the tree toad and the elephant, the giraffe and the kangaroo, the wolf and the lamb should all be the descendants of a common ancestor? Yet these and all other creatures must be blood relatives if man is next of kin to the monkey" (Bryan 1922, 103).

Above all, Bryan hated Darwinism for its moral implications. "Darwinism leads to a denial of God. Nietzsche carried Darwinism to its logical conclusion and it made him the most extreme of anti-Christians. . . ." (Bryan 1922, 123). This, for Bryan, is all served up in an unholy stew of Teutonic power hunger. "As the war progressed I became more and more impressed with the conviction that the German propaganda rested upon a materialistic foundation. I secured the writings of Nietzsche and found in them a defense, made in advance, of all the cruelties and atrocities practiced by the militarists of Germany." The consequences of Darwinism are appalling. "Nietzsche's philosophy would convert the world into a ferocious conflict between beasts, each brute trampling ruthlessly on everything in his way" (124). The moral damage has been incalculable. "Nietzsche died hopelessly insane, but his philosophy has wrought the moral ruin of a multitude, if it is not actually responsible for bringing upon the world its greatest war." Parenthetically one might add that this, to say the least, is an exaggeration. One is not quite sure whether to praise Bryan for seeing that German philosophy feeds into such aggressive thinking at least as much as does the evolutionary theorizing of Charles Darwin; or to criticize him for being unaware that, notwithstanding debts, ultimately one of the greatest critics of the volkish, pan-Germanic movement, so beloved of the military, was none other than Friedrich Nietzsche.

It was surely the plan of their Sovereign God that Fosdick and Bryan, good Presbyterians both, should clash. In a letter to the *New York Times*, Fosdick wrote: "When, therefore, Mr. Bryan says, 'Neither Darwin nor his supporters have been able to find a fact in the universe to support their hypothesis,' it would be difficult to imagine a statement more obviously and demonstrably mistaken." Adding: "So far as the general outlines of it are concerned, the Copernican astronomy itself is hardly established more solidly" (Fosdick 1922, 262). Bryan had his revenge. At the 1923 General Assembly of the Presbyterians, the Fundamentalists gained the upper hand, passing a resolution querying Fosdick's general commitment to the faith, and this led to his resignation. It didn't hurt that the hugely wealthy John D. Rockefeller picked up the tab for the new ecumenical Riverside Church, in New York City, of which Fosdick became the pastor.

War Comes

Expectedly, when the Second War was declared, most of the churches stepped smartly into line, supporting the conflict. In Germany, until the war, elements of both Protestantism and Catholicism had been strongly opposed to the Third Reich. On commencement of war, Hitler eased up on his religious opponents and they, both Protestant and Catholic, raised no opposition to the invasion of Poland and beyond, and many stressed support. The Protestants: "We unite in this hour with our people in intercession for our Führer and Reich, for all the armed forces, and for all who do their duty for the fatherland." In the same mode, Catholics: "We appeal to the faithful to join in ardent prayer that God's providence may lead this war to blessed success for Fatherland and people" (Conway 1968, 234). The same was true elsewhere. Dutifully, the Church of England took up the role as national cheerleader and such happenings as the Luftwaffe bombing of Coventry Cathedral, on the night of November 14, 1940, became a national symbol of suffering and resistance—as indeed it still is. Obviously, in Britain, whether one was a pacifist or not, the regime did not call for the moral opposition demanded of those under the Nazis. Britain did not have or need its heroic opponents like Pastors Niemöller and Bonhöffer. There were those nevertheless who stood firmly against the blind enthusiasm characteristic of the First War.[1] Bishop Bell of Chichester always sought to separate the aims of the Church of England from those of the secular state. "It is the function of the Church at all costs to remain the Church . . . It is not the State's spiritual auxiliary with exactly the same ends as the State" (Chandler 2016, 152). This led to contact during the war with the German resistance (through his friendship with Bonhöffer) and a plea—entirely ignored—that the allies aim only for the defeat of the Nazis rather than the crushing of the German people.

There were even some who, on theological grounds, argued against Britain's declaring war on the Nazis. In the middle years of the twentieth century, Gertrude Elizabeth Margaret Anscombe (1919–2001) was a formidable force in British philosophy. A convert to Catholicism during her

[1] To be fair, generally one does not sense the blind enthusiasm seen earlier. In part, this was because Hitler and the Nazis were so vile, one had no need to whip up false emotion to fight them. It was more a necessary job to be done. In part, although the losses were great, in the West—not the East—they were far from the horrendous figures of the Great War. Again, there was no need to whip up false emotion to tide one over the truly unacceptable. I do not discount the costs of this conflict—at sea, in the air, on land. Look no further than the sobering trilogy, by Rick Atkinson (2002, 2007, 2013), on the fighting in North Africa, in Sicily and Italy, and in France and the Low Countries.

student years, its ethical teaching influenced her strongly. Expectedly, as one who came to maturity just as the Second World War was about to begin, she was always interested in the moral issues surrounding war. Putting her discussion in the Catholic context, she never had much time for pacifism. Like Niebuhr, she seemed to concede that "Absolute pacifists" like Mennonites could take that position with respect, but, like Niebuhr again, those active in society were not just wrong but positively evil. They are so against the shedding of any blood that they cannot see that acting in the cause of justice is one thing and the harming of innocents is quite another. "Now pacifism teaches people to make no distinction between the shedding of innocent blood and the shedding of any human blood. And in this way pacifism has corrupted enormous numbers of people who will not act according to its tenets" (Anscombe 1981, 57–58). Continuing: "Pacifism and the respect for pacifism is not the only thing that has led to a universal forgetfulness of the law against killing the innocent; but it has had a share in it."

This said, Anscombe set stringent just war constraints on the waging of war and, notoriously, as a student at the beginning of the Second World War, spoke out strongly against taking up arms against the Germans. She, with a co-author (Norman Daniel), argued vehemently that the cause was not just. As philosophers, they began by laying out the ground rules. In particular, discussing the morality of war brings on the idea of natural law, which "is the law of man's own nature, showing how he must choose to act in matters where his will is free, if his nature is to be properly fulfilled. It is the proper use of his functions; their misuse or perversion is sin." Adding: "To those that believe in God it will rightly appear that His law, the eternal law, has its reflection in the ordered activity of Creation, that 'law of nature' which is the truth of things" (Anscombe and Daniel 1939, 72). They list seven conditions for a just war: just occasion (violation of rights); declaration by lawful authority; upright intention; right means of conduct; war the only possible means to right wrong; reasonable hope of victory; and probable good outweighs probable evil. They agreed that, with respect to the declaration of war in 1939, certain conditions were met. There was a just occasion, meaning that there was a violation of rights, namely the invasion of Poland. The war was ordered by lawful authority, the leaders of the democratically elected British parliament. There was a reasonable hope of victory. And all things considered, war was the only recourse to put things right. Peaceable means were simply no longer an option.

However, the aims are not just. No one really cared that much about Poland. The real reason for going to war was to beat Germany back to the status quo, and given the unfairness of the Versailles Treaty that means the reason was to reinstate something that was in itself wrong. "Our policy, it might be said, is incomprehensible, except as a policy, not of opposing German injustice, but of trying to preserve the status quo and that an unjust one" (74). The means are going to be unjust, for they will undoubtedly involve attacks on civilians. Nor can one use the excuse that in modern warfare everyone in a war is a combatant. Apart from the young or the old or the sick or whatever, people going about their business are not thereby fighting. "The civilian population behind an army does not fulfill the conditions which make it right to kill a man in war. Civilians are not committing wrong acts against those who are defending or restoring rights." Agreeing that civilians "are maintaining the economic and social strength of a nation," yet adding, what others would no doubt have considered less than well supported, "that is not wrong, even though that strength is being used by their government as the essential backing of an army unjustly fighting in the field" (78). For Anscombe (and Daniel), the principle of double effect—a Thomistic argument that bad effects can be excused if they come contingently as the side consequences of actions for the good—doesn't help here either. Blockades and (obliteration) bombing set out deliberately to harm civilians, and that is impermissible. Apart from anything else: "No action can be excused whose consequences involve a greater evil than the good of the action itself, whether these consequences are accidental or not" (78). Using the kind of reasoning that so swayed Fosdick, they invoked the authority of Aquinas. "The force used must be proportional to the necessity." Adding: "As we have seen, our government does intend to do that which is unlawful, and it is already blockading Germany with intent to starve the national life. The present war is therefore wrong on account of means" (79).

Finally, there is the question of the possible evil effects of the war. The writers dismiss worries of the Nazis going free. "Already men are talking of Germany as a pariah nation; they are already saying that she must henceforward be kept down and never allowed to become powerful again." They wrote of this as "insane determination, which is as foolish as it is immoral." Continuing, "after the war, what prospects have we but of greater poverty, greater difficulties, greater misery than ever, for a space; until just another such war will break out" (81).

Leave the discussion here. After the revelation of the horrors of Bergen-Belsen and of Auschwitz, talk like Anscombe's sounds somewhat hollow, not to say naïve. Perhaps it is unfair to judge her too harshly in the light of what was to come. Ask our important linking question. Are we within the framework, or have we lost sight of our characterization of the essence of Christianity and of how this speaks to the problem of war? Humans as the creation of a good God, and yet fallen into sin. We can be saved, but only through the mercy of God, for which we have no right or justification. War comes because of human sin and is ever with us. We must engage or not engage in war against this background. Progress in any real and lasting sense is chimerical.

Although not conventional just war theorists, both Niebuhr and Barth in important respects think within this Augustinian framework. We are sinners. In *The Nature and Destiny of Man*, remember based on lectures given in 1939, we get a very detailed discussion of original sin. It is linked firmly to Paul's declaration (Romans 5:13) that it is something we inherit thanks to the sin of Adam. We learn that we will sin, freely, for all that it is something we cannot avoid. "Original sin, which is by definition an inherited corruption, or at least an inevitable one, is nevertheless not to be regarded as belonging to his essential nature and therefore is not outside the realm of his responsibility. Sin is natural for man in the sense that it is universal but not in the sense that it is necessary" (Niebuhr 1941, 242). From this, it is predictively certain that war will come about. Yet, for all that war is an evil, it could be God's will that we wage it—although for Niebuhr it is all rather complex, for it seems as though we have to sin (go to war) to fulfill God's ends. "To the end of history the peace of the world, as Augustine observed, must be gained by strife" (Niebuhr 1932, 335).

This, almost expectedly, got Niebuhr into some frightening discussions, particularly about the obliteration bombing by the allies in the Second World War—where you simply bomb everyone and everything, including civilians, in your striving to win. Catholic theologians spoke eloquently against this. One can never do such evil in order to achieve good. "It is fundamental in the Catholic view that to take the life of an innocent person is always intrinsically wrong, that is, forbidden absolutely by natural law. Neither the state nor any private individual can thus dispose of the lives of the innocent" (Ford 1944, 272). Bishop Bell joined this discussion with a strong condemnation of obliteration bombing. The Church "must not hesitate, if occasion arises, to condemn the infliction of reprisals, or the bombing of civilian populations, by the military forces of its own nation" (Towle 2010, 34–35). This was much to the chagrin of Churchill and the Archbishop of York, among others. This may well have cost him the Archbishopric of Canterbury. Across the Atlantic, in continued denial of just war theory—in applying it, judgments "are influenced by passions and interests, so that even the most obvious case of aggression can be made to appear a necessity of defence" (Niebuhr 1941, 283)—Niebuhr was stuck with saying that fighting in a war is always sinful anyway, so obliteration bombing is really more of the same. "Once bombing has been developed as an instrument of warfare, it is not possible to disavow its use without capitulating to the foe who refuses to disavow it. No man has the moral freedom to escape from these hard and cruel necessities of history. Yet it is possible to express the freedom of man over the necessities of history" (Niebuhr 2015a, 655). In the end, Niebuhr said that the bomber is a sinner, but then so are we all—"the Lord's Supper is not a sacrament for the righteous but for sinners"—and fortunately the Kingdom of God is "the peace of divine forgiveness, mediated to the contrite sinner who knows that it is not in his power to live a sinless life on earth."

Whatever, there is not much place here for progress or the perfectibility of humankind. Certainly, war in that sense is not going to lead to human improvement. What of those who went in another direction? What of the pacifists? Although he broke with Augustine on the Christian morality of waging war, William Jennings Bryan accepted the essential framework of the Bishop of Hippo. He didn't believe in evolution, at least not of humans. He believed in Adam and Eve and the fall. He believed in sin and he believed that it is only through the grace of God that salvation is possible. Going to Tennessee and appearing in the Scopes Monkey Trial was as much a statement against the modern-day philosophy of unbridled, secular progress as it was in favor of Genesis taken literally (Larson 1997).

Fosdick is in major respects more complex and more interesting. Let no one deny his authentic Christian faith or that he believed his savior died for his sins on the cross. The question is about how one interprets all of this and in particular how one interprets all of this in the context of war. Fosdick believed in evolution. He did not believe in a literal Adam and Eve. Hence, although he knew we were all sinners, he was not committed to the traditional notion of original sin. For this reason, although as an American Protestant he clearly put everything in a Providential framework—it is God, God, God—he also put everything in a developmental framework.

> There are multitudes of Christians, then, who think, and rejoice as they think, of the Bible as the record of the progressive unfolding of the character of God to his people from early primitive days until the great unveiling in Christ; to them the Book is more inspired and more inspiring than ever it was before. To go back to a mechanical and static theory of inspiration would mean to them the loss of some of the most vital elements in their spiritual experience and in their appreciation of the Book. (Fosdick 1922, 720)

This means that we humans can make for real improvement, and that is our God-given duty. "Development does seem to be the way in which God works. And these Christians, when they say that Christ is coming, mean that, slowly it may be, but surely, his will and principles will be worked out by God's grace in human life and institutions, until "he shall see of the travail of his soul, and shall be satisfied."" In theological terms, this means that Fosdick was more postmillennial—trying to make for paradise here on earth—than premillennial—waiting for God's return knowing that no genuine improvement is now possible. It is not ridiculous therefore—nor is it theologically forbidden—to hope for an end to war. In respects, therefore one sees glimmers of possibility—to be explored before the end of this book—that the Christian and Darwinian positions on war might achieve some sort of harmony. Note though that it is harmony not integration. Fosdick is first and foremost a Christian. He feels comfortable drawing on his Christian tradition. For instance, although, in tandem with his appeal to the Gospels, he is no just war theorist, he agrees with Augustine's central point that the means must always be proportional to the end. No end could ever justify the slaughter of the Great War. Here, if anything, he is on stronger grounds than Niebuhr. There are things we Christians must never do. And if that seems unrealistic, so be it. Everything in human life and institutions "is being worked out by God's grace." Perhaps Pastor Fosdick was premillennial after all!

8 | From Hitler to UNESCO

THE TIME SPAN OF THIS chapter, roughly from the end of the First World War to the end of the Second World War, was the most vital (since the days of the *Origin* and the *Descent*) in the history of the development of Darwinian evolutionary theory as a mature, professional science (Provine 1971; Ruse 1996). Pick up where we left off in Chapter 1.

The Darwin-Mendel Synthesis

The biggest scientific problem faced by Darwin was that of heredity. If natural selection is to work, then favorable variations must be passed on to future generations. Which means that they must not be washed out or diluted away by the process of breeding. As one astute commentator noted, in a very Victorian example, if a white man arrived on an island of natives, however good he might prove at reproducing, it is not likely that in three or four generations the population will have evolved significantly lighter skin (Greg 1868). Darwin tried to speak to this problem with his "provisional hypothesis of pangenesis," supposing that little gemmules or particles are given off from all over the body, carried to the sex cells, and then transmitted to future generations (Darwin 1868). This way, because ultimately things are particulate, one might expect to see features endure and, at the same time, Darwin could speak to his Lamarckian yearnings in that if a feature in a body is changed—say the blacksmith's arms get stronger through working at the forge—these changes will be reflected in the gemmules and passed on.

No one was very much impressed by this, although equally no one had much else to offer (Olby 1963). Things hung fire until the beginning of the twentieth century, when the work of the Moravian monk Gregor Mendel was discovered and it was realized that a viable theory of heredity was within grasp (Bowler 1989). Before long—thanks particularly to the labors of Thomas Hunt Morgan and his associates in New York City—the "classical theory of the gene" was articulated, seeing the units of heredity (genes) as physical things lined along the chromosomes in the nuclei of cells (Allen 1978). It was not immediately obvious that Mendelian genetics and Darwinian selection theory were complements—two sides making the whole picture—rather than contradictories and rivals, but by the second decade of the new century people were starting to realize this. Since Darwinian selection is a mechanism that works only in groups—one set beating out another set—the key move was in generalizing thinking of the gene focused on individuals—what features does a parent pass on to a child?—to thinking of the gene focused on the population—what ratios of different kinds of gene (alleles) hold from generation to generation? The breakthrough here was the formulation of the so-called Hardy-Weinberg Law (named after its co-discoverers), which basically said that if there are no external influencing factors—like selection, or mutation (spontaneous change from one form of a gene to another), or immigration or emigration—then gene ratios stay the same. Counterintuitive, at least to many, this means that even a very small ratio of one kind of gene or allele will persist in a (large) population, indefinitely. Now, with this equilibrium law—a kind of biological equivalent to Newton's First Law of Motion (bodies stay at rest or in uniform motion unless acted upon by a force)—"population geneticists" could start to find their theorems and to build their systems (Provine 1971).

Three figures are crucial. In England, there were Ronald A. Fisher and J. B. S. Haldane. In America, there was Sewall Wright. History shows that while there was formal equivalence, there was quite significant difference in the thinking about the nature of evolutionary change, especially between the more influential Fisher and Wright (Provine 1986). The former had training as a classical physicist and always thought of genes in a population rather like gas molecules in a container. He saw natural selection working on the whole and thus moving populations gradually from one set of ratios to another set of ratios. It was all a bit like the synchronized hand waving that goes around sports stadia. The latter, who (as an employee of the US Department of Agriculture) worked intensively on the breeding patterns of short-horn cattle, was more focused on change in small populations, where chance could overcome expected patterns of variation. Calling this "genetic drift," Wright argued that large populations fragment, that chance changes dominate in the fragmented parts, and then they recombine where selection takes over and favors the better forms of one kind of subpopulation over others (Wright 1931, 1932). Introducing the powerful metaphor of an "adaptive landscape," with different genes (alleles) in three-dimensional space—two dimensional over an area and then the third dimension denoting adaptive highs and lows—Wright spoke of his "shifting balance theory of evolution" with equilibrium being disturbed by external factors, fragmentation, drift, combination and selection, and then a new time of equilibrium. He owed more of this to Herbert Spencer than to Darwin; not surprising, for in America especially, Spencer had always a devoted following (Ruse 2004).

With the theory in place by the early 1930s, the naturalists and empiricists moved in and, as it were, started to put biological flesh on the mathematical skeleton (Ruse 1996). In Britain, the key person was E. B. Ford, who founded the school of what he called "ecological genetics." Working on fast-breeding organisms like Lepidoptera and snails, he and his associates and students—notably H. D. B. Kettlewell, D. H. Sheppard, and A. J. Cain, not to mention Fisher who was ever the *éminence grise*—did sterling work showing the power of selection in populations. Most famously, they spoke to the ongoing issue of industrial melanism—the way that organisms turn darker for camouflage reasons when smoke dirties the background vegetation (Majerus 1998). In the United States, the key person was the Russian-born geneticist (he trained with Morgan) Theodosius Dobzhansky (1937), who did sterling work on Drosophila, the fruit fly, both in the laboratory and in the field. Although initially very empathetic to Sewall Wright's shifting balance theory, he soon found that selection was much more significant in natural populations than was drift, and he produced a stunning, ongoing series of publications showing just that (Lewontin et al. 1981). Always something of a scientific entrepreneur, Dobzhansky not only had many students of his own—some of whom, notably Richard Lewontin (1974) and Francisco Ayala (2009), were to make the move to the molecular world with panache and great success—he brought in biologists from other areas to fill out neo-Darwinism (to use the British name) or the synthetic theory of evolution (to use the American name). Particularly important were the German-born systematist and naturalist Ernst Mayr (1942), the paleontologist George Gaylord Simpson (1944), and the botanist and geneticist G. Ledyard Stebbins (1950).

To restate the claim of the *Prolegomenon*, by mid-century or a little later—certainly by 1959, the hundredth anniversary of the *Origin*—we now had a selection-based, fully functioning paradigm of evolutionary studies (Ruse 1996). Let there be no doubt that for all of my focus on Darwinism as a religion, this was a normal scientific enterprise, with the marks of such an enterprise. In Britain there was the journal *Heredity*, in America the journal *Evolution*. Evolutionists moved from fringe positions in natural history museums and agricultural stations to proper university jobs. Fisher was now a professor at Cambridge, Haldane at University College London, Wright at Chicago, and the others were all firmly ensconced in academia. Grants were applied for and received. The Nuffield Foundation in England was a real cornucopia of funds—much was made of the medical benefits of knowing about the spread and change of genes in populations. Butterflies are great model organisms! After the Second World War, America poured money into the National Science Foundation, and evolutionists found their begging plates well filled. All of this has continued non-stop down to the present. Recently, there have been exciting moves toward "evolutionary medicine," thus blowing away Thomas Henry Huxley's skepticism about whether Darwinism could cure a pain in the belly (Nesse and Williams 1994; Gluckman, Beedle, and Hanson, 2009).

Darwinism as Religion

Darwinism as a religious alternative to Christianity did not go away. It took something of a breather after the First World War, as the biologists worked on building a professional science of evolutionary thinking, but equally the more popular side to the science developed and really took off in a spectacular way. It is very much with us today. Mention has been made already of the voluminous writings of the English biologist Richard Dawkins, author of the *Selfish Gene* (1976) among others; and of the American paleontologist (the late) Stephen Jay Gould, author of *Ever Since Darwin* (1977a) among others. Mention has been made also that they were into a lot more than straight science and that that "lot more" moved in the direction of religion is given away by their obsessions and publications. Notoriously Dawkins is the author of the *God Delusion* (2006); Gould of *Wonderful Life* (1989) and *Rocks of Ages* (1999), as well as other writings, especially his monthly column ("This view of life") of the magazine *Natural History*.

Taking a breather in the years after the Great War does not mean that the Darwinian equivalent to Christianity ever vanished entirely. The essence of the Christian side is that of Providence—tainted as we are by original sin, we can do nothing without the grace of God—and the essence of the Darwinian side is that of social or cultural or technological progress—unaided, we ourselves can make for a better tomorrow. Every one of the evolutionists was an enthusiast for progress, some were really quite fanatical and obsessive on the subject, and most had little hesitation in introducing this idea into their regular science. In other words, as noted there was no Popperian nonsense about science being value free. Or, at the least, there was no nonsense about value-free science being kept from needs of Darwinism, evolution as religion. Thus Fisher, to take a British example, and extracted right from his classic *The Genetical Theory of Natural Selection*: "Entropy changes lead to a progressive disorganization of the physical world, at least from the human standpoint of the utilization of energy, while evolutionary changes are generally recognized as producing progressively higher organization in the world" (Fisher 1930, 296). Let there be no question about the winners. "I have never had the least doubt about the importance of the human race, of their mental and moral characteristics, and in particular of human intellect, honour, love, generosity and saintliness, wherever these precious qualities may be recognized." Interestingly, Fisher was a deeply committed Anglican, even giving sermons in his college chapel, but it was his Christian religion that was molded to fit his Darwinian philosophy, rather than conversely. "In the language of Genesis we are living in the sixth day, probably rather early in the morning, and the Divine Artist has not yet stood back from his work, and declared it to be 'very good.' Perhaps that can only be when God's very imperfect image has become more competent to manage the affairs of the planet of which he is in control" (Fisher 1947, 1001). This, frankly, does not sound very Augustinian to me.

In the United States, showing that he thought evolutionism was a threat to his own religious position, Niebuhr went after Edwin Grant Conklin, one of the most influential biologists in the country in the first half of the twentieth century. With good reason, for Conklin was almost laughably Spencerian in his thinking about evolution. "Life itself, as well as evolution, is a continual adjustment of internal to external conditions, a balance between constructive and destructive processes, a combination of differentiation and integration, of variation and inheritance, a compromise between the needs of the individual and those of the species." The metaphysics was there too. "Progress is the product of the harmonious correlation of organism and environment, specialization and co-operation, instinct and intelligence, liberty and duty" (Conklin 1921, 87). The population geneticists were in the same camp. Not only was Wright an enthusiast for Spencer—which in itself was probably a major reason why he preferred group selection over individual selection—he was also keen on the thinking of the Frenchman Henri Bergson. In his seminal paper on the shifting balance theory, Wright wrote: "the present discussion has dealt with the problem of evolution as one depending wholly on mechanism and chance. In recent years, there has been some tendency to revert to more or less mystical conceptions revolving about such phrases as 'emergent evolution' and 'creative evolution.' The writer must confess to a certain sympathy with such viewpoints philosophically" (Wright 1931, 155). Bergson had said (in his *Creative Evolution*) that we humans might be thought of as "the reason for the existence of the entire organization of life on our planet" (Bergson 1911, 92).

The Evolutionists on War

Agree then that the backbone of progress continues to stand ramrod straight up. What then of war? The older men had tangled with the First War. Not all were fighters. Fisher's eyesight (to his great regret) precluded service. He lost a son in the Second World War. Wright tried to enlist in 1916 (remember America did not enter the war until 1917) but was underweight. He went on an eating spree but by the time he had some success, the war was nigh over. The British geneticist Lancelot Hogben, later to gain fame as the author of the best-selling *Mathematics for the Millions*, was a conscientious objector. He ended up in Wormwood Scrubs. "Apart from communal work in the laundry one morning a week and about twenty minutes' walk around the yard in single file for morning exercise in fine weather, we were almost continuously in solitary confinement with a daily allotted task of sewing pre-cut strips of canvas to make mail bags" (Hogben 1998, 62). Interestingly, in the Second World War, he felt able to work on military statistics.

Julian Huxley, who was lecturing in the United States at Rice University in Houston, came home and joined military intelligence. In the Second World War, he got himself into hot water when, in the early years of the European war and before America joined in, he gave a talk in New York City urging the country to drop its neutrality and come alongside the British against Hitler (Baker 1978). Most interesting and, typically, flamboyant of them all was Haldane (Clark 1968). With Scottish connections, he served in the Black Watch, a regiment known by the Germans as the "ladies from hell," partly on account of their ferocity and partly because they wore kilts. Haldane himself, labeled by his chief (Haig) "the bravest and dirtiest officer in my army," rather enjoyed the whole experience and reveled in killing Germans. Indeed, other soldiers did not altogether relish his enthusiasm, which was liable to attract fire to them. He suffered injuries and also took time out to help his father, the eminent physiologist J. S. Haldane, on experiments to make better protections from gases. Then and later he had an almost masochistic insistence on first trying out things on his own body. In the late 1930s, he went to Spain as an observer and saw rather more of aerial bombing than is comfortable.

In the Second World War, Haldane ranted against the government in the Marxist newspaper the *Daily Worker.* Secretly, he was helping the government on topics like the body's resistance to carbon dioxide—a matter of some urgency because of the threats to well-being from submarine life. (There was a terrible accident just before the war, when the submarine HMS *Thetis* had sunk with great loss of life.) With Haldane, one had always to look at the big picture and to discount his inbuilt urge to shock. In 1925, he actually published a book, *Callinicus*, defending chemical warfare. When one digs into the details, the reasoning makes (let us say) for uncomfortable reading. Supposedly, people of color—negroes, Indians, and the like—have greater resistance to mustard gas than white people. Therefore, use the colored people (meaning this in the generic sense) as forward shock troops, and get on with the job from there. Fewer killed overall and quicker decisions too. "It would not much upset the present balance of power; Germany's chemical industry being counterpoised by French negro troops. Indians may be expected to be as nearly immune as negroes" (Haldane 1925, 50). One might add that Haldane was not alone in his views about the tolerance of different races toward mustard gas. In the Second World War, America tested out the gas on blacks and Japanese, leading to years of horrific discomfort (not to mention blank, authoritarian denial) for the survivors.

Later Haldane was to redeem himself somewhat with his thinking on defenses against aerial bombing. He had little time for those who followed Stanley Baldwin who said "the bomber will always get through." Adding: "The only defence is in offence which means you have to kill more women and children quickly if you want to save yourselves" (Anon. 1932, 7). Based on his experiences in Spain, Haldane sounded like one of our Catholic priests on the subject of obliteration bombing. "I do not think that there can be any reader who does not feel in the marrow of his or her bones that this is a wicked policy. This is equally true for absolute pacifists and for those who have been or are members of the fighting forces. And like all immoral statements, it is untrue. There is a defence other than murder." Haldane then went on to discuss ways in which shelters should be built in the face of bombing as well as various evacuation possibilities. His position was one of emotion as well as reason. "Air raids are not only wrong. They are loathsome and disgusting. If you had ever seen a child smashed by a bomb into something like dirty rags and cat's meat you would realize this fact as intensely as I do." We must do something. "Above all, I hope, in spite of the terrible international situation, that this book will prove unnecessary, and that the people of Britain will never see what I have seen in Spain" (Haldane 1938, 11). Faint hope, I am afraid, both in Britain and in Germany.

Eugenics

You might object that this is all very well, but really we need to know what the evolutionists said in the name of their theory of evolution about war. This objection is not entirely well taken. Someone like Haldane, whether talking of the use of gas in attack or shelters in defense, is appealing to science as the way forward. In other words, he is setting his discussion in the context of technological progress, which is very relevant to our discussion. Turn now directly to evolution and war. Start with the question of eugenics. What of the fear that war takes the brightest and best, thereby impoverishing the population and pointing to decline? We certainly see elements of this thinking between the wars. Fisher was most concerned about the overbreeding of the inadequate against the better-quality members of society, even going so far as to promote family allowances for the middle and upper classes only, as a spur to retraction. Ever the ardent eugenicist, he expressed the war-is-dysgenic fears (in his classic of 1930). "The hero is one fitted constitutionally to encounter danger; he therefore exercises a certain inevitable authority in hazardous enterprises, for men will only readily follow one who gives them some hope of success." Continuing: "It is undeniable that current social selection is unfavourable to heroism, at least in that degree which finds it sweet as well as proper to give one's life for his country. Any great war, will reveal I believe, a great fund of latent heroism in the body of almost any people, though a great war must sensibly diminish this fund" (Fisher 1930, 265).

One had to be fairly guarded in what one was going to say in the years after the Great War. One hardly wanted to proclaim too loudly and publicly that all of the good genes are now gone and all that is left is second class and cowardly. Sir Arthur Keith, with his enthusiasm for the group, was somewhat of an exception to this. Although eugenics flourished in the years we are considering, much of it was more negative eugenics, concerned with cutting down the breeding of the undesirables, than much in the positive line (Kevles 1985). With massive state-supported programs of sterilization, in both Europe and the Americas, that was enough. Certainly, there were biological voices speaking out against eugenics and by implication the deleterious effects on a nation's bloodstock because of war. J. B. S. Haldane, on his way to Marxism, had little time for worries about such things. Being Haldane, he could not resist a snide comment: "Curiously enough eugenic organizations rarely include a demand for peace in their programmes, in spite of the fact that modern war leads to the destruction of the fittest members of both sides engaged in it" (Haldane 1934, 9). His basic position, however, was that it is culture that matters, not the genes. "I have yet to come across any evidence whatsoever that there has been any advance in the intrinsic factors making for intelligence in Europeans during the last 50,000 years" (8). It is culture and technology where we find the causal action. "Progress as far as I can see has been due to the substitution of one type of production by another, and in so far as the new social organization has been stable, the progress has been of a fairly permanent character." Of course, there is biological progress, but not such as is of interest to us. "It is no doubt desirable that man should evolve in certain directions, but such evolution is a quite different thing from social progress. It may be that there is a limit to the social progress possible without further evolution, but before such a conclusion is proved a good many experiments will have to be made; and the statement that the limit of progress has now been reached need not be taken seriously except as an expression of conservatism in the speaker" (8).

What of the people we saw wrestling with the biology of the First War? Kellogg, remember, worried much about the deleterious genetic effects of war, but in the end decided that it was necessary to fight the Germans. After the war, as we saw, his thinking on gene-culture coevolution had changed, with a move much closer to that to be expressed by Haldane. Humans started off as animals. "But by the nature of our physical evolution, which gave us speech and the possibility of recording our traditions, and gave us a special development of mind rather than better claws and teeth with which to carry on our struggle for existence, our general course of evolution diverged importantly from that of the other great animals in that it moved toward development on a basis of the mutual aid principle rather than the mutual struggle principle" (Kellogg 1919, 97). Altruism. As insects honed in on its virtues biologically, so humans honed in on its virtues culturally. Thinking back to the earlier evolutionists and their various stands on the biological versus cultural causes, Kellogg combined the two, thinking that we changed biologically over time, but now this kicks in and makes possible cultural evolution. As already noted, although a Christian, Kellogg had no hang-ups about original sin making real change impossible. "We have a social or traditional inheritance as well as a physical inheritance. And it is by conscious modification of our environment and education that we can determine the character of this all-important influence on our lives" (98). One is rather glad that this very decent man did not live long enough to see the full horrors that were yet to come.

German Thinking

We have also encountered the post-war General Friedrich von Bernhardi. It is true that having noted the sacrifice made by the high-quality Germans—"a great part of our fine officers' corps lies on the battle fields; another has been forced by wounds and sickness to leave its beautiful profession" (von Bernhardi 1920, i)—without denying that there must be some, he was a little more circumspect about the overall genetic effects. However, essentially the old warhorse was unchanged, blaming bad strategy on Germany's defeat and vehemently protesting the terms of the Versailles Treaty. It won't and can't work (Krischel 2010). He was, unfortunately, right about this. Although there were many other factors involved—hyperinflation following the French occupation and then, later in the decade, the knock-on effects of the American Depression—the soil was being well fertilized for the germination and growth of Adolf Hitler (Evans 2003).

We must take care here. There is today a cottage industry of Creationists and fellow travelers who, having failed to show that Darwinism is epistemologically or metaphysically flawed, are now faulting it on moral grounds, namely that it led straight to Adolf Hitler, the Third Reich, the Holocaust, and the overall Götterdämmerung of the 1940s (Gasman 1998; Weikart 2004). Prima facie, there is something to all of this. If you look at Charles Darwin's *Descent of Man*, published in 1871, you will find some very Victorian ideas about the races. At the other end, if you look at Adolf Hitler's *Mein Kampf*, you will find some passages that do seem to draw on Darwinian theory. "Those who want to live, let them fight, and those who do not want to fight in this world of eternal struggle do not deserve to live" (Hitler 1925, 1, chapter 11). Yet, as always, the truth is more nuanced. I don't think you can or should say definitively that there are no links. Apart from anything else, something had to lead to Hitler and the Nazis, and if you eliminate Luther (and his horrendous anti-Semitism) and eliminate Darwin and eliminate—well, you know the tune—you end up with no causes at all leading to the wicked movement that overtook Germany in the 1930s. I would be very surprised if the anti-Semitism of Christianity and the racism of the nineteenth century had no causal role. However, before you rush to conclude that the Creationists are correct and that there are significant links between Darwin himself and Hitler, consider a number of points (Richards 2013).

First, reemphasize the Darwinian opposition to slavery (Desmond and Moore 2009). The members of the Darwin family were fanatical, anti-slavery campaigners. In the early part of the nineteenth century, when the young Darwin was growing up, this was the family obsession. It rubbed off on him. On the voyage of the *Beagle*, on the question of slavery in South America, he had a huge row with his captain, Robert Fitzroy. During the American Civil War, he was a strong supporter of the North, precisely because of the slavery issue. (Many in Britain supported the South because of the links with the cotton trade.) The *Descent*, for all that it did reflect the concerns of a middle-class, Victorian gentleman, was no clarion call to racial superiority. Darwin was explicit that when the races met and (as so often was the case) the non-Europeans suffered, it came not from intellectual or social superiority but because non-Europeans caught the strangers' diseases, fell sick, and died.

Second, think of what we have learned so far. While it is true that many used Darwin's ideas to promote social policies, and that some used them to promote aggression—General Friedrich von Bernhardi for a start and apparently just about every member of the German general staff that Vernon Kellogg met in Belgium—never forget there were others who promoted very different ideas. The co-discoverer of natural selection, Alfred Russel Wallace, was an ardent socialist and feminist in the name of Darwinism. The Russian Prince Petr Kropotkin argued for anarchy in the name of Darwin. And so on, before we get to William James and Vernon Kellogg himself. Very much like Christianity, you can argue that Darwinism supported a plethora of contradictory positions. This being so then, very much like Christianity, one might ask just how genuine and important was the support being offered. There was a propaganda value, true. Genuine links are another matter

Third, when you turn to Hitler himself, the story is murky. To put the matter politely, he was not a well-educated man (Kershaw 1999). There is no evidence he studied Darwin's writings or much about them. At most, he was picking stuff up off the street or from the barroom or from the doss house where he lived in Vienna before the war. When you look at *Mein Kampf* in more detail, the story seems less straightforward.

Just before the apparently Darwinian sentiments quoted above, Hitler wrote: "All great cultures of the past perished only because the originally creative race died out from blood poisoning." What he is really on about are the Jews. Darwin would have been appalled at such a connection. So in the end, reject Creationists and fellow travelers as bad theologians, bad philosophers, and bad scientists. While you are at it, reject them as bad historians. Charles Darwin was not to blame for the First World War. He was not to blame for Adolf Hitler. At least, inasmuch as he was, so also are Grimms' fairytales.

Behavior

It is important not to whitewash evolutionists. Hitler and the Nazis generally were not that keen on evolution. Apart from anything else, it proclaims the unity of humankind—we are the same as Jews under the skin! However, there were evolutionists, Darwinian evolutionists, who were happy to support and embellish the Nazi race views. One was the Austrian, future Nobel Prize winner Konrad Lorenz—a pioneer in the field of "ethology," a kind of proto-sociobiology where one studies the behavior of animals from an evolutionary perspective. Turning from his scientific work of the 1930s, when he worked with graylag geese—it is to him that we owe the notion of imprinting—Lorenz became a booster for the political powers. "I'm able to say that my whole scientific work is devoted to the ideas of the National Socialists." Lorenz bought in completely to all of the Nazi guff about clean living and the open air and getting back to our peasant roots. "The Nordic movement has been emotionally guided, from time immemorial, against the 'domestication' of human beings." We have two options: going on as we are or turning around and moving toward what we once were. "For a biological sensibility, no doubt can exist as to which of the two ways is the true way of evolution, the way 'onward and upward!' " (Kalikow 1983, 66, quoting Lorenz "Durch Domestikation," written in 1940). Again, though, one should be careful about how Darwinian is any of this. Lorenz had a father with very strong eugenical views. There is also the example of Lorenz's fellow Prize winner, the Dutch ethologist Niko Tinbergen, who ended up as a prisoner of war of the Germans. There is certainly no simple causal chain from Charles Darwin to Nazi-supporting ethology. One might add that one wonders how seriously to take a writer who, in an essay on the dangers of domestication, uses the figure of Socrates as an example of what happens when you get too far from your simple, pastoral roots.

There was no connection between the study of behavior and the holding of right-wing or fascist views. Across in America at the University of Chicago, Quincy Wright, the political scientist brother of the geneticist Sewall Wright, was running a major study on war, culminating in his own massive 1,552-page-long book—*A Study of War*. He showed the influence of his brother, seeing evolutionary biology at the root of things.

> Among primitive peoples before contact with civilization warfare contributed to the solidarity of the group and to the survival of certain forms of culture. When population increased, migrations or new means of communication accelerated external contacts. The warlike tribes tended to survive and expand; furthermore, the personality traits of courage and obedience which developed among the members of these tribes equipped them for civilization. (Wright 1942, 1287–1288)

Perhaps not entirely consistently—across his vast territory, not an easy task one imagines—Wright did see biology involved also in the reluctance to fight. He spoke of "the general existence of hereditary patterns against interspecific killing in all organisms." Very much in line with the thinking of his brother, Wright invoked a group-selective mechanism. "The generally accepted theory that organic evolution tends to shape individuals with behavior patterns not unfavorable to survival of the species provides an explanation for this pattern" (92). Wright was a technical advisor at the Nuremberg trials and a big supporter of the United Nations, thinking that we can work toward some kind of universal peace.

Warder Clyde Allee—departmental colleague of Sewall Wright—was likewise giving a very different, non-Nazi-friendly spin to things (Mitman 1992). As a Quaker, he had little time for original sin in the sense of something that taints us because of the behavior of others long ago. He would have stressed rather the idea of the Inner Light, "that of God in every man," seeing us all with the potential for good. "I am the light of the world: he that followeth me shall not walk in darkness, but shall have the light of life" (John 8:12). Allee argued that "colony life arose from the consociation of adult individuals for cooperative purposes." Continuing: "In terms of human society, this view would stress the importance of the gang, rather than the family, as a preliminary step in the evolution of the social habit. *It is important to note that the gang cuts across family lines in its formation*" (Allee 1927, 391). With the emphasis on selection working at the group rather than the individual or family level, Allee—like the Wright brothers—was not precisely following Darwin himself. But his overall stance did enable Allee to open the way to a society envisioned as beyond war. This was much the line he took in a book, *The Social Life of Animals*, published just before the Second World War. He was not very keen on the idea that humans have an "inherent, instinctive drive toward war" (Allee 1938, 188). He seemed to think that war comes because of external factors, notably overpopulation pressures. Allee was not convinced, however, that these must always be a problem, for already population growth—for unknown reasons, possibly contraception—is slowing quite dramatically. Nor, if this means an end to war, should we as Darwinian evolutionists regret this fact. All the old eugenical worries, David Star Jordan is cited as an important authority, are trotted out. "Personal selection, so far as it exists in modern warfare, selects the individual to be killed or wounded because he is physically or mentally superior to those who are left at home" (204). We must work therefore toward a kind of refurbished League of Nations, where all peoples come together to prevent strife and war. "There seems to be no inherent biological reason why man cannot learn to extend the principle of co-operation as fully through the field of international relations as he has already done in his more personal affairs" (217).

We have in Allee a very instructive variant reading of the Darwinian approach to war. He had no time for Augustinian sin, and equally he had no time for man the innately violent. In fact, in an address given during the Second War, he blamed evolutionists by name—particularly Thomas Henry Huxley and Herbert Spencer—for using evolution to encourage warlike thoughts. "There will be some truth in the accusation, for biological science is not wholly free from war guilt. This is not only because we have been the inventors of tools for mass destruction but because we have been responsible for giving interpretations to some aspects of Darwinian theories of evolution that provide a convenient, plausible explanation and justification for all the aggressive, selfish behavior of which man is capable" (Allee 1943, 524). In Allee's view, we can and should reject this kind of thinking and transcend war—it is immoral and, as bad, it is biologically stupid, and it has no basis in evolutionary thinking. "The biological support for this fatalistic view regarding, among other things, the inevitableness of intra-species human conflict, is now opposed by strong evidence which indicates that the idea of a ruthless struggle for existence is not the whole, or even the major contribution of current biology to social philosophy and social ethics." You might object that Allee a Christian, a man who was a Quaker and thus a pacifist, should not be taken as an exemplar of Darwinian thinking. Things are a little more nuanced. Because he was a Quaker, he did not accept the Augustinian take on original sin, and so he was free to go in other directions. "There are both egoist and altruistic forces in nature, and both are important" (521). Which is the more important? "Under many conditions, the cooperative forces lose. In the long run, however, the group-centered, more altruistic drives are slightly stronger." Keep this thinking in mind. It nicely points to the overlap and possible link between (one strain of) traditional Christian thought about human nature and (one strain of) traditional Darwinian thought about human nature.

Julian Huxley

Turn now to the figure, in our time period, in the context of our discussion—Darwinian evolutionary theory and thinking about war—whose role cries out for recognition. No one, starting with Huxley himself, would have said that Julian Huxley was a great scientist; but, in a way, he is even more significant. He used his talents and energies to become an important cultural figure, running the London Zoo, starring on radio and later on television, and being the first director general of UNESCO—it was he who insisted on the S for Science being added to that organization (Waters and van Helden 1992). Above all, Huxley wrote books including the popular work that put new developments in Darwinian thinking firmly on the screen—*Evolution: The Modern Synthesis* (1942).

Start with the fact that Huxley, a nonbeliever, thought of himself nevertheless as being very religious, and—as we saw in Chapter 2—set out deliberately to formulate a science-based (in major respects a Darwinian-based) alternative to Christianity. Moreover, he put this in the very terms of our discussion—Providence versus progress. "The release of God from the anthropomorphic disguise of personality also provides release from that vice which might be termed Providentialism. God provides for the sparrow, we are told; how much more for man? And so this beneficent power will always provide" (Huxley 1927, 18). Prominent in Huxley's thinking is the theme that Providential thinking leads to complacency and acceptance of the unsatisfactory. It is all a question of Providence versus progress. For good measure, Huxley threw in the blame for war on Christianity. "The European War, the further one penetrates into its history and causes, seems to have been inevitable; but war itself is not therefore necessarily inevitable." But it did happen because we humans were inadequate. Religion didn't help. "God did help bring about the War—but God in our impersonal and not absolute sense; a god of which no mean part was human ideals not yet purged of selfishness, greed, and combativeness, a god still partly tribal" (16).

If humans are the endpoint, the apotheosis of evolution, how did Huxley think we progressed up to us? Arms races! In the spirit of Darwin, it was he who developed the idea, before the Great War, making the case in a little book, *The Individual in the Animal Kingdom* (1912). Then, after the war, restating the case in *Animal Biology*, a textbook published jointly with Haldane. Deeply influenced by the progressionist thinking of the French philosopher Henri Bergson, stressing the competition between evolving lines leading to improvement, obviously inspired by the prewar competition between the British and German navies, Huxley gave a graphic description of an arms race couched in terms of naval, military technology. "The leaden plum-puddings were not unfairly matched against the wooden walls of Nelson's day." Now however, obviously having in mind the then-huge competition between Britain and Germany, "though our guns can hurl a third of a ton of sharp-nosed steel with dynamite entrails for a dozen miles, yet they are confronted with twelve-inch armor of backed and hardened steel, water-tight compartments, and targets moving thirty miles an hour. Each advance in attack has brought forth, as if by magic, a corresponding advance in defence." Likewise in nature, "If one species happens to vary in the direction of greater independence, the inter-related equilibrium is upset, and cannot be restored until a number of competing species have either given way to the increased pressure and become extinct, or else have answered pressure with pressure, and kept the first species in its place by themselves too discovering means of adding to their independence" (Huxley 1912, 115–116). Eventually: "it comes to pass that the continuous change which is passing that through the organic world appears as a succession of phases of equilibrium, each one on a higher average plane of independence than the one before, and each inevitably calling up and giving place to one still higher."

I don't think anyone, even Julian Huxley, would want to say that arms races in themselves are valuable—hurling great chunks of metal across the ocean to kill other human beings doesn't seem much like value in any absolute sense. However, once you have introduced relative value—big bombs good, bigger bombs better—it is hard not to slide into absolute value. Science—Darwinian biology—does not insist that conflict and war are good things. Indeed, science as such tries by its very nature to be value neutral. However, with the introduction of thoughts of progress you have introduced values, and it is hard to escape the conclusion that, in Huxley's world, competition, arms races, and war are or were of some real value. They led to us! "It should be clear that if natural selection can account for adaptation and for long-range trends of specialization, it can account for progress too. Progressive changes have obviously given their owners advantages which have enabled them to become dominant." There is even an air of inevitability, certainly predetermination, about the emergence of humans at the top. "One somewhat curious fact emerges from a survey of biological progress as culminating for the evolutionary moment of the dominance of *Homo sapiens*. It could apparently have pursued no other general course than that which it has historically followed" (Huxley 1942, 333).

To put the icing on the cake, as it were, where did Huxley stand on the course of cultural evolution and its future prospects for society and the threat of war? Here we turn to a little essay that Huxley wrote in 1948 when he became director general of UNESCO, a little essay one might add that waxed enthusiastic about such issues as birth control and so upset some of his conservative and religious sponsors that his four-year term was reduced to two. The whole point of UNESCO as a major organ of the United Nations was to move the world on beyond war. The preamble to the charter is explicit. "We the peoples of the united nations determined to save succeeding generations from the scourge of war, which twice in our lifetime has brought untold sorrow to mankind." Huxley had no doubt about how this was to be achieved. UNESCO! "Its main concern is with peace and security and with human welfare, in so far as they can be subserved by the educational and scientific and cultural relations of the peoples of the world. Accordingly its outlook must, it seems, be based on some form of humanism" (Huxley 1948, 5). Continuing: "It must also be a scientific humanism, in the sense that the application of science provides most of the material basis for human culture, and also that the practice and the understanding of science needs to be integrated with that of other human activities."

Huxley incidentally was a great enthusiast for mega-scientific and technological enterprises and had a year or two earlier written a very enthusiastic book about the virtues of the Tennessee Valley Authority, bringing electricity to the rural south of the United States (Huxley 1943). The point about such enterprises is that they do not exist in their own right, but for the betterment of humankind. This is why scientific humanism cannot "be materialistic, but must embrace the spiritual and mental as well as the material aspects of existence, and must attempt to do so on a truly monistic, unitary philosophic basis." Hence, the humanism of UNESCO "must be an evolutionary as opposed to a static or ideal humanism. It is essential for Unesco to adopt an evolutionary approach. If it does not do so, its philosophy will be a false one, its humanism at best partial, at worst misleading" (5).

In Julian Huxley, we could not have a better exemplification of the traditional world picture I am ascribing to the Darwinians. Explicitly, he is pushing a secular religious picture in opposition to Christianity; he is for progress and against Providence; he thinks that Christianity is at least partially responsible for war; say what you like, at some level he has to agree that war in the past was a good thing because it led to that all-valuable species, humankind; and he thinks that all of this points the way forward to a society of enriched culture and altruism, where the chances of and need for war are significantly reduced.

The Inklings

It was also a picture that was going to upset the Christians as inevitably as the emergence of humans in Huxley's world. Although it was to be the 1950s and later that the fantasy tales of the so-called Inklings—a group of Oxford chums including C. S. Lewis and J. R. R. Tolkien—were to be published and start upward toward their incredible success, already in the 1930s they were writing and preparing the way (Zaleski and Zaleski 2015). Two things stand out above all. First, these men had lived and fought through the First World War and it marked them forevermore. Although (because of his age) Lewis was a latecomer to the war, he fought in the trenches and was wounded (by friendly fire that killed two of his comrades). Tolkien fought in the Battle of the Somme and later was invalided out of the front line because of trench fever. He suffered on and off from illness until finally he was demobilized. Later in life, he stressed always that it was the First War rather than the Second that obsessed his creative imagination.

Second, these men were ardent Christians, even though they were not uniform in their faith. Lewis converted from atheism to High Anglicanism—a somewhat ironic fact given how today he is the darling of American Evangelicals. Tolkien followed his mother in a deep commitment to Catholicism. Another member, Owen Barfield, a solicitor and a major influence on both Lewis and Tolkien (as well as somewhat of an influence on T. S. Eliot), was a follower of Rudolf Steiner (whom he judged a much greater man than Goethe). The religious commitments came through in their writings. As did worries about wars, past and present. In 1940, Lewis gave a remarkably bellicose talk against pacifism, something that would have done credit to a Great War Anglican bishop. Paul and Augustine are touted as better guides than Jesus to the true meaning of Christianity, the authority and the aura of the state (meaning England) over the individual are explained and endorsed in detail, and any speakers for the opposition are quite ignored (Lewis 1976).

Christian themes figure quite (some would say obtrusively) strongly in the fantasy writings of Lewis—Aslan the lion sacrificing himself for Edmund the sinner and the subsequent resurrection—less obviously in Tolkien perhaps, but still strongly there. "The Lord of the Rings is of course a fundamentally religious and Catholic work; unconsciously so at first, but consciously in the revision. That is why I have not put in, or have cut out, practically all references to anything like 'religion,' to cults or practices, in the imaginary world. For the religious element is absorbed into the story and the symbolism" (Carpenter and Tolkien 2000, letter of 1953 to Robert Murray S.J.). The very fact that the Inklings were fantasy writers (not to mention professors of English rather than, say, biochemistry) points to their unease with science and technology and a yearning for a simpler world, a rural world (like the Shire) where sophisticated machines are unknown and people are in more direct contact with nature. This led readily to an instinctive dislike of the proselytizing scientists of their day, a leader among whom was Julian Huxley (Bud 2013). Combining an atheistic materialism with a yearning for vitalism—both heresies to good Christians—he was a ready target. C. S. Lewis's Space Trilogy—above all the final volume, *That Hideous Strength* (published in 1945)—had huge amounts of propaganda about the perils of scientific thinking, with materialism, vitalism, and the degradation of human nature ever a danger. Set around a place known as the National Institute for Coordinated Experiments, the corrupting influence of philosophies of progress were all too apparent. "Dreams of the far future destiny of man were dragging up from its shallow and unquiet grave the old dream of Man as God. The very experiences of the dissecting room and the pathological laboratory were breeding a conviction that the stifling of all deep-set repugnances was the first essential for progress" (Lewis 1945, 203).

Tolkien, likewise, had the philosophy Huxley represented firmly in his sights. The treatment of war—a dominating theme in his *Ring Trilogy* (Tolkien 2005)—is very Augustinian (probably via Aquinas and other medieval figures). On the one hand, through just war theory it is made clear that it is legitimate and proper for men of goodwill to take up arms against the evil foe. At the same time, however, the conduct must be legitimate.

With other Catholics, Tolkien was dismayed at aspects of the British and American conduct in the Second World War and he decried strongly the demonizing of the enemy—"The Germans have just as much right to declare the Poles and Jews exterminable vermin, subhuman, as we have to select the Germans: in other words, no right, whatever they have done" (letter to his son Christopher, 1944). On the other hand, never ever forget how inevitably vile and awful are war and its aftermath. Any talk of war as a necessary element toward the greater good of progress is obscene. Some of the most powerful writing clearly represents those dreadful days on the Somme.

All of this is set in the overall Christian picture of humans as fallen creatures, tainted with original sin, where we can (and must) indeed make all our efforts to fight against evil, but where ultimately we alone are bound to fail and salvation comes only through the grace of God. Crucial to the Ring story is that, at the final moment, when Frodo and Sam have arrived at the point where the Ring can be destroyed, Frodo hesitates and falls—he puts the Ring back on his own finger, lusting for the ultimate power that it represents. Matters are rectified only when the pathetic but dangerous Gollum tears off Ring and finger and falls to his death, and the destruction of the Ring, in the fire far below. We are not saved through our own merit, but through grace. In the end, it is made clear that peace has arrived again, but constant vigilance is needed for evil always waits once more to rear its ugly head. This rests on a subtle point of Catholic theology: evil is never a positive thing but always an absence of good. This is why it is an ongoing threat and can never be eliminated. Good carries within itself the possibility of its negation.

Leave matters here. No one has ever as fully articulated the Darwinian position as did Julian Huxley. Yet always, as he criticized Christianity for its thinking on war, in turn he and his fellows had their Christian critics.

9 | The Bomb and Vietnam

BECAUSE OF HIS AUTHORIZING THE bombing of Hiroshima and Nagasaki, in the 1950s Elizabeth Anscombe led a campaign at Oxford against the awarding of an honorary degree to Harry Truman. "In the bombing of these cities it was certainly decided to kill the innocent as a means to an end. And a very large number of them, all at once, without warning, without the interstices of escape or the chance to take shelter, which existed even in the 'area bombing' of the German cities" (Anscombe 1957, 64). The end can never justify the means, not even when the end is ending the war against the Japanese. Many would not have agreed. Paul Fussell, the literary-historian author of *The Great War and Modern Memory*, a truly great work on the writings in and about that conflict, was a combat soldier in Europe when the Second World War ended against Germany. He writes movingly at the relief that the dropping of the Bomb meant to one such as he, until then destined to go to the other end of the world and invade Japan. However, at a more general level, Anscombe was not alone in her worries.[1] Although it might be expected from her as a Catholic, she was also not alone in the second half of the twentieth century in showing more philosophical appreciation for traditional arguments than were shown previously. Perhaps it was the abject

---

[1] On a personal note, it was at this time (around 1960) that I first brushed in a personal way against the issues being discussed in this book. As an undergraduate, over several Easters I marched—singing lustily—from Aldermaston, where the British bombs were made, to Trafalgar Square in central London.

Men and women, stand together,
Do not heed the men of war.
Make your minds up now or never,
Ban the bomb for evermore.

Once, I heard the very aged Bertrand Russell speak. A great thrill. I did not then join him in getting arrested for sitting in protest outside 10 Downing Street, the Prime Minister's residence. That was a privilege reserved for people of far-higher status than I.

failure—often absence—of earlier ad hoc argumentation that spurred a greater reaching back to intellectual roots.

Protestant Concerns About the Bomb

Paul Ramsey, a long-time teacher of Christian ethics at Princeton, was a Methodist who wrote on just war theory in the 1960s and later. At once, we can infer a number of things about his interests and his writings. First, he was a Protestant. That meant, for all that he was returning to roots, not being a Catholic, Ramsey's intellectual engagement with traditional just war theory would be at a distance, as it were, and this might well (as it does) show in his use of the theory. Second, although Ramsey was trained by Reinhold Niebuhr's brother, Richard, he was not a Lutheran or Calvinist. Methodism, an off-branch of Anglicanism, accepts original sin, but tempers it with a belief in "prevenient grace." Unlike Calvinists particularly, who stress predestination and God's total sovereignty, Methodists believe (in the tradition of Jacob Arminius and in parallel with Kant) that God has given us free will to choose freely what He offers. Much influenced by Pietism, and hence allied with the Radical branch of the Reformation, Methodism stresses love of one's neighbor. The beatitudes are important as one reaches out to those in need and in want of help. It is not pacifist like Quakerism, but hatred and violence are very much against its spirit. Like Kant, another deeply influenced by Pietism, the ends can never justify the means. The Categorical Imperative states flatly that we are always to treat others as ends in themselves, and never as tools. Third, Ramsey was writing at the time of the Cold War when the Bomb—the A bomb and then the H bomb—dominated all thinking and was a matter of truly vital concern. At the push of a couple of buttons, could America and Russia tomorrow simply end life here on Planet Earth? The theology is a theology of love. Never, ever, could one or should one wantonly kill the innocent. "No ethics—least of all Christian ethics—gives us leave to kill another man's children directly as a means of weakening his murderous intent. Preparation to do so—if that is the real and the only object of our weapons—is intrinsically a grave moral evil" (Ramsey 1961, 11–12).

In building and making his case, Ramsey had (let us say gently) a somewhat idiosyncratic view of the history of war theory. We have noted general opinion that the turn from absolute pacifism came with the adoption, by the Roman Empire, of Christianity as the official religion. Pacifism could never function in such a situation. One looks therefore for a break between the thinking of early Christians and post-Constantine Christians. For Ramsey, this was bad history. It was much more a question of having the same theology in a different context. "Christians simply came to see that the service of the real needs of all the men for whom Christ died required more than personal, witnessing action. It also required them to be involved in maintaining the organized social and political life in which all men live. Non-resisting love had sometimes to resist evil" (xvii–xviii). Note how this means that love is and continues to be central. It also means that, as with other Protestants like Barth and Niebuhr, traditional natural law theory is going to be secondary, if even that, and biblical inspiration is going to be all-important. Just war theory will not stand alone, as it would for a Catholic, but will be interpreted through this particular Protestant lens.

> The just-war theory, in both its aspects, requiring the use of force and limiting the use of force, is nothing more than an application for the supreme standard for Christian living as this is received in Protestant Christianity, namely, "Everything is quite lawful, absolutely everything is permitted which love permits, everything without a single exception. . . . Absolutely everything is commanded which love requires, absolutely everything without the slightest exception or softening." Therefore, the Christian is permitted to use force, nay, even positively obliged to do so; and he is at the same time required to limit the use of it, nay, his faith permits him to limit it in situations where men are hard pressed to do wrong, vainly imaging that in this, God's word, some good may come of it. (190–191)

Given its roots in natural law theory, we find that Ramsey is not that interested in reasons for making war (*jus ad bellum*)—proper authority, for instance. He is going to be much more interested in the conduct of war (*jus in bello*). This reflects in his two major concerns, namely (in Anscombe's terms) "only the right means must be used in the conduct of the war" (discrimination of combatants and noncombatants) and "the probable good must outweigh the probable evil effects of the war" (what Ramsey called "proportionality"). One should say that this focus on the conduct of war, rather than reasons for war, was surely also a function of the threat of nuclear weaponry. Going to war, when you have the Bomb, is not that different in principle from going to war without the Bomb, but in practice it is very different. When you have a weapon of mass destruction, things are changed, for then you do at once come face to face with issues to do with innocents, not to mention horrendous problems about proportionality. "I say simply that any weapon whose every use must be for the purpose of directly killing noncombatants as a means of attaining some supposed good and incidentally hitting some military target is a weapon whose every use would be wholly immoral" (162). Issues of double effect don't even come in here, because you are killing people as a means in its own sake, not accidentally as a means for some other goal. This before you even get to proportionality issues.

Does this mean then that the Bomb should always be banned? Not so fast! Ramsey said that "megaton weapons would always destroy military objectives only incidental to the destruction of a whole area; and that in the very weapon itself, its use, its possession, or the threat to use it, warfare has passed beyond all reasonable or justifiable limits" (163). About such a bomb then he qualifies himself: "the *Grenzmorality* [morality on the border or a knife-edge] of its merely military possession and use depends on whether in fact, now or in the future, there are any conceivable circumstances in which it would have importance against military targets against which less powerful weapons would not serve as well. In the fluidity of historical events and changes in the concentration of political and military power perhaps this cannot be entirely ruled out as a possibility" (163–164).

This is a slippery slope, because now Ramsey faced the problem of nuclear deterrence. What about building a Bomb to scare the other side, even saying that one will not strike first but in the event of an attack will use the Bomb to cause so much destruction that the attack in the first place proves simply not to be worth it? (This is a Pearl Harbor situation. The Japanese caused much destruction, but bringing on the wrath and power of the United States proved in the long run that it was a foolish thing to have done.) Here, Ramsey gets caught in a very unsatisfactory dilemma. On the one hand: "To press the button in counter-retaliation will also be the most unloving deed in the history of mankind, only exceeded by those who for the sake of some concern of theirs, cause the little ones to stumble and fall into hell" (170). On the other hand, Ramsey seems to realize that making threats about such a possibility may not only occur but in some sense be needed. He seems to slip over to the realism of Niebuhr. We get the private morality/public morality division in full force. Individuals *qua* individuals should never behave immorally. "But there should be statesmen who themselves are quite clear as to the immorality of obliteration warfare (and as well the wrong of deterring evil by readiness to do the same thing) who are still willing to engage in negotiation directed to the end of limiting war to justifiable means and ends through a period of time in which they may have to defer their nation's repentance" (12). Adding: "Thus, it may be that a "just war" Christian may sometimes find himself supporting a nation's preparations for unjust warfare—as there have been pacifist Christians in public office who have been willing to vote for a military appropriation" (12).

A number of balls are being juggled in the air and one senses that one at least has just fallen to the ground. Not that this would have worried the then Archbishop of Canterbury, Geoffrey Fisher. His comment on the nuclear situation was that "for all I know it is within the providence of God that the human race should destroy itself in this manner. There is no evidence that the human race is to last forever and plenty of Scripture to the contrary effect" (Toynbee 1959, 43). Adding: "The very worst the Bomb can do is to sweep a vast number of People from this world into the next into which they must all go anyway." Fisher, like Temple before him, had been the headmaster of a private ("public") boys' school, Repton (the same as Temple). I suspect that all of us who have had extended interactions with adolescents feel this way sometimes, but as a general theological inference it does seem a bit extreme.[2]

More seriously, one might note as a codicil that, in the Archbishop's statement, there is an uncomfortable echo of the loudly promoted views of a branch of Christianity very far, socially and geographically, from the prelate. Many American evangelicals are not only premillennialists, they are also dispensationalists (Vox 2017). The brainchild of a nineteenth-century member of the Plymouth Brethren, John Nelson Darby, it is believed that the course of the world (of short, biblical dimension) is interrupted by God clearing the decks, as it were. The expulsion from Eden, Noah's Flood, Tower of Babel, and so forth. The dispensations are the ages before the event—the time in Eden being the first. We are living in the last. After the Second War, this kind of thinking grew exponentially in American evangelical circles, often combined with extreme opposition to evolutionary theory, which (with reason) was thought of as attacking the very foundations of "premillennial dispensationalism" (Whitcomb and

[2] About one of his charges at Repton, Michael Ramsey, Fisher's much-beloved successor to the Archbishopric, Fisher cautioned the then-prime minister, Harold Macmillan: "I have come to give you some advice about my successor. Whomever you choose, under no account must it be Michael Ramsey, the Archbishop of York. Dr Ramsey is a theologian, a scholar and a man of prayer. Therefore, he is entirely unsuitable as Archbishop of Canterbury. I have known him all his life. I was his Headmaster at Repton." To which Macmillan responded: "Thank you, your Grace, for your kind advice. You may have been Doctor Ramsey's headmaster, but you were not mine" (Hennessy 2001, 250).

Morris 1961; see also Ruse 1982, 1988). A major spur was the perceived, growing threat of a nuclear confrontation. If such occurred, so the interpretation went, we would surely have the final, biblically forecast dispensation, Armageddon.

Perhaps unfairly linking to Fisher, what is striking is that so many who were convinced by this somewhat idiosyncratic theological picture were almost enthused by the prospect. God was going to do this anyway, people are going to go to heaven or not, so why not welcome it and get on with it? Giving inversely proportional attention to the quantity of literature produced, I will pass on merely noting that some of the theology was a little more conventional, and to a person premillennial dispensationalists are completely and utterly committed to an Augustinian belief in original sin. It was the misdeed of Adam that brought to an end the first dispensation and gave God reason and need to return as the Christ. Naturally, this was taken as opening the way for the necessity of Christians to wage war, even of the worst kind. The prospect of end times gave added force to such thinking. Above all, beware of false prophets. Heed the words of Hal Lindsey, author of *The Late Great Planet Earth* (1970), thirty million copies in the first two decades and counting: "Look for some limited use of modern nuclear weapons somewhere in the world that will so terrify people of the horrors of war that when the Antichrist comes they will immediately respond to his ingenious proposal for bringing world peace and security from war" (185). UNESCO!

Just War Theory

The Vietnam War was waged mainly in the second half of the 1960s and the first half of the 1970s. It was a war into which America entered to prop up the South of the country against the communist North. Despite America's overwhelming military might and willingness to sacrifice tens of thousands of young men (sixty thousand dead) in an alien country far from home, it was an aim that was not successful. This gave a new urgency to discussions about war and morality. From a stream, overnight we had a flood. We had now a huge number of non-Christians joining the debate, starting with the most influential of them all, Michael Walzer, a Jew, author of *Just and Unjust Wars*. Of course, whether you subscribe to a non-Christian religion or to no religion at all, whether you are a believer or an agnostic or an atheist, you have every right to join the discussion and to make use of Christian theory as you find it useful. About so major a topic as war and its moral standing, we all benefit from serious discussion, from whatever quarter. Moreover, as has been stressed before, Christian thinking owes much to non-Christian ideas. Natural law theory is the case in point. However, we should expect to find, if not shifts, changes of emphasis. Take Paul Ramsey and his stress on the love commandment. A nonbeliever might well think that love is important but not give it the primacy that, as part of his Christian commitment, Ramsey gives it. Analogously discussing pacifism, a nonbeliever will not work against the eschatological background of the Christian pacifist, thinking that ultimately whatever happens down here on earth, God will right things in the hereafter.

Keep these points in mind, and remember that the overall aim is a comparison between Christian and Darwinian ideas, rather than between Christians and Darwinians per se. In this light, because of its importance, let us look first at Walzer's thinking, but then move sharply back into the Christian world. By his own admission, although he immersed himself in the just war literature and history, military history really informed Walzer's thinking when he wrote *Just and Unjust Wars*. He had never been a soldier himself, and in his reading he "focused on histories, memoirs, essays, novels, and poems because I wanted the moral arguments of my book to ring true to their authors—and to the men and women about whom they were writing" (Walzer 1977, 336). It is this indeed that makes his work so gripping and immediate—again and again we start or end with a case study, illustrating or supporting (or denying) a point. It is done brilliantly, and I empathize strongly with this approach. I do not see how one could think seriously about proportionality without thinking of concrete situations like the Somme—so many young men from Britain and its Empire dead on the first day. However, Walzer's approach does call for more construction of argument than usual in these sorts of discussions. Although Walzer himself is helpful here for he has said that, above all, he is proposing an argument about the moral standing of war. Right or wrong? "The argument is twofold: that war is sometimes justifiable and that the conduct of war is always subject to moral criticism. The first of these propositions is denied by pacifists, who believe that war is a criminal act; and the second is denied by realists, for whom "all's fair in love and war": inter arma silent leges (in time of war the laws are silent)" (3).

As far as pacifism goes, Walzer offers a counterargument, not so much criticizing pacifism—remember, he does not have the tools of the Christian here—as making the case for a just war. It all depends very much on community and the legitimacy and value of the relations between its members. If a state is just a collection of people, then nothing follows. "But most states do stand guard over the community of their citizens, at least to some degree: that is why we assume the justice of their defensive wars. And given a genuine 'contract,' it makes sense to say that territorial integrity and political sovereignty can be defended in exactly the same way as individual life and liberty" (54). Walzer then goes on to say that other states may legitimately join in the defense. "The victim of aggression fights in self-defense, but he isn't only defending himself, for aggression is a crime against society as a whole. He fights in its name and not only in his own. Other states can rightfully join the victim's resistance; their war has the same character as his own, which is to say, they are entitled not only to repel the attack but also to punish it" (59). Walzer stresses however that the right of a state to defend itself is the only reason for war—meaning here presumably including the right of a state to come to the defense of another state. In the First World War, Britain coming to the aid of Belgium, when it was invaded by Germany. "Preventative wars, commercial wars, wars of expansion and conquest, religious crusades, revolutionary wars, military interventions—all these are barred and barred absolutely, in much the same way their domestic equivalents are ruled out in municipal law" (72). One wonders what George Washington and his pals would have thought about that.

As far as realism is concerned, Walzer takes the same tack, namely pointing out that we simply don't think it is true. Distinguishing between *jus ad bellum*, reasons for war, and *jus in bello*, conduct in war, Walzer argues that in both cases it is clear that there are moral constraints governing war and its waging. Just starting a war because we want to—for conquest or religion, for instance—is simply immoral. "We know the crime because of our knowledge of the peace it interrupts—not the mere absence of fighting, but peace-within-rights, a condition of liberty and security that can exist only in the absence of aggression itself" (51). Likewise, there are rules of conduct. Suppose a soldier shoots another soldier on the opposite side. "Assuming a conventional firefight, this is not called murder; nor is the soldier regarded after the war as a murderer, even by his former enemies" (128). By extension, it is not murder if the soldier on the other side kills you. Talk of murder only comes in when soldiers "take aim at noncombatants, innocent bystanders (civilians), wounded or disarmed soldiers. If they shoot men trying to surrender or join in the massacre of the inhabitants of a captured town, we have (or ought to have) no hesitation in condemning them" (128). The subtext here is the My Lai massacre in Vietnam, in 1968, where US Lieutenant Calley led his men to kill up to 500 unarmed villagers, not to mention gang rape and mutilation of bodies.

Predictably, Walzer's treating of all soldiers as being moral equals has occasioned much discussion. Does one really want to say that there should be no condemnation of a troop of SS soldiers, even if they are scrupulous in following the rules of war? One might say that there are some things that no one should do or be and this is one of them.

Exploring Implications

As a political scientist rather than a theologian or philosopher, Walzer's style is less one of laying out a theoretical framework and then commenting on it, and more one of giving examples and commenting on these, although obviously with some notion of underlying structure. Whether working from a Christian perspective or not, he is always deeply thoughtful and at times very helpful and insightful. Take for instance the contentious question of whether it was moral or wise for the allies to have demanded unconditional surrender of the Germans in the last world war. Ramsey argues that this was wrong. "The love or agreement that constitutes a people, and informs their justice and the laws of their peace, may justly be defended; but for intrinsic reasons this may not be done in such a manner and for such ends that the whole people and its entire welfare are ventured and at stake on the outcome." Demands for unconditional surrender are therefore wrong. A lot here obviously depends on what one means by "unconditional surrender." If it means that the victor is going to put to the sword the whole population of the losing side, this is grossly immoral. Walzer points out rightly that no one intended this of Germany. To do this would have been to lower oneself to the level of the Nazis, or lower if that be possible. However, it is legitimate, when facing an enemy run by such an organization as National Socialism, an organization that clearly breaks any moral norm that it is possible to conceive of, to demand that the running of the country be given over entirely to the victors and the leaders of the losers be subject to prosecution, even if this means they are put to the sword. It is the moral vileness of Hitler and his henchmen that is the foremost issue. Demands for unconditional surrender were not as such to prevent future happenings but to express "a collective abhorrence, a reaffirmation of our own deepest values. And it is right to say, as many people said at the time, that the war against Nazism had to end with such a reaffirmation if it was to end meaningfully at all" (117).

Less successful is Walzer's discussion of blockades. In the First World War, both Britain and Germany tried to starve the foe into submission through blockading the incoming supply of materials, especially food. The Germans tried to do this through submarines; the British through conventional ships on the ocean's surface. The British won the battle. Germany crumpled and collapsed. One could try to excuse the actions of both sides by the principle of double effect. One is not trying to starve the noncombatants. One is trying to starve the combatants and those who aid them, not to mention depriving the combatants of raw materials needed to continue the fight. In fact, with respect to Germany, there were not so many instances of death through outright starvation, but up to half a million civilian deaths did occur from the effects of disease on weakened bodies. The British defended their actions, saying that the War Cabinet "had planned only a 'limited economic war,' directed, as the official history has it, 'against the armed forces of the enemy.' But the German government regrettably maintained its resistance, 'by interposing the German people between the armies and the economic weapons that had been leveled against them and by making the civil populace bear the suffering inflicted'" (173).

Walzer comments witheringly: "the sentence invites ridicule." This still leaves open the real reason—obviously to beat Germany into submission—and its moral status. Agree that it was immoral to starve the civilian population. Remember, Germany was doing the same to Britain. Were Britain's leaders to order off the navy against German ships, simply spending their time picking up survivors from British ships that had been sunk by German submarines? Either one reverts to realism—morality doesn't come into things here—or one plays Ramsey's dodge of saying that public behavior is not private behavior and thus sometimes people have to do bad things for the public good. Or, a third option, one takes Niebuhr's strategy of saying that in the public domain, private morality simply doesn't apply. One has to be a "Christian realist." None of these solutions seems entirely happy on Walzer's terms. Nor does just rolling over and letting the other side win. Morally upright losers are still losers.

The point now is less to criticize and more to say that Walzer truly reinvigorated just war theory, albeit doing so from a non-Christian perspective. He pointed people to a framework within which they could think about war, and it is little wonder that he spawned a full and vigorous area of inquiry by philosophers and others. Among those were people thinking explicitly in a Christian context.

Specifically, there was "The Challenge of Peace: God's Promise and Our Response: A Pastoral Letter on War and Peace by the National Conference of Catholic Bishops," issued in 1983. The bishops put their Christian stance right at the beginning of the discussion. "At the center of the Church's teaching on peace and at the center of all Catholic social teaching are the transcendence of God and the dignity of the human person. The human person is the clearest reflection of God's presence in the world; all of the Church's work in pursuit of both justice and peace is designed to protect and promote the dignity of every person." We start here. "God is the Lord of life, and so each human life is sacred; modern warfare threatens the obliteration of human life on a previously unimaginable scale. The sense of awe and 'fear of the Lord' which former generations felt in approaching these issues weighs upon us with new urgency.' "

What then does this mean in concrete terms with respect to peace and war? The bishops were concerned to acknowledge and respect, and make room for, the pacifist. However, Christians are in the business of defending peace, in the right sense, against external threats and violence. This is a given. Where discussion arises is over the exact means of defending such peace. These are the big moral issues. Stressing how they wrote not just on the basis of biblical teaching but also the tradition and authority of the church, they added: "Pope Pius XII is especially strong in his conviction about the responsibility of the Christian to resist unjust aggression: A people threatened with an unjust aggression, or already its victim, may not remain passively indifferent, if it would think and act as befits a Christian. All the more does the solidarity of the family of nations forbid others to behave as mere spectators, in any attitude of apathetic neutrality." Working in an acknowledged Augustinian mode, the bishops endorsed the belief that, although war is unfortunate, given our fallen nature it will be with us always. "Augustine was impressed by the fact and the consequences of sin in history—the 'not yet' dimension of the kingdom. In his view war was both the result of sin and a tragic remedy for sin in the life of political societies." Saying also that the idea of a totally and everlasting peaceful human society is a fantasy, a utopia. If you think this way, you are simply out of touch with human nature—either through ignorance or because you are too scared to face up to the truth or because you have nefarious motives. There is truly no escape. Human nature, as it is, makes war that from which we cannot escape.

The bishops' overriding concern was with nuclear weapons. To this end, they referred explicitly to traditional just war theory, in the context of starting war (*jus ad bellum*) making much of the concept of proportionality. This "means that the damage to be inflicted and the costs incurred by war must be proportionate to the good expected by taking up arms." Adding that this is something that governs the conduct of war as well as the decision to go to war in the first place. It was precisely these issues that led earlier bishops to condemn the Vietnam War. The end simply did not justify the cost. This brings the bishops specifically to nuclear war: "Under no circumstances may nuclear weapons or other instruments of mass slaughter be used for the purpose of destroying population centers or other predominantly civilian targets." This gives rise to an unacknowledged paradox. On the one hand, retaliation is, if not ruled out, very strongly circumscribed. "Retaliatory action whether nuclear or conventional which would indiscriminately take many wholly innocent lives, lives of people who are in no way responsible for reckless actions of their government, must also be condemned." On the other hand, deterrence seems at least a possibility: "In current conditions 'deterrence' based on balance, certainly not as an end in itself but as a step on the way toward a progressive disarmament, may still be judged morally acceptable" (Pope John Paul II, Message to UN Special Session on Disarmament, #8, June 1982). It is not clear how one might have deterrence without the real threat of massive retaliation. If the enemy knows you are a paper tiger, you are left with no defense. "Specifically, it is not morally acceptable to intend to kill the innocent as part of a strategy of deterring nuclear war." One suspects that realists might have something to say here, especially since savvy opponents would know how to render the principle of double effect worthless. "The location of industrial or militarily significant economic targets within heavily populated areas or in those areas affected by radioactive fallout could well involve such massive civilian casualties that, in our judgment, such a strike would be deemed morally disproportionate, even though not intentionally indiscriminate."

The strength of coming at an issue from a Christian perspective, a Catholic perspective, is that not only does one have a honed theory of war from which to work, but one has the general moral tradition of the church giving authority to what one argues and proposes. The difficulty with taking such an approach is that one is liable to find oneself facing problems simply not anticipated when the theory was first articulated or even later when it was refined. Nuclear weapons so change the area of discourse that old distinctions like those between killing innocents as an unwanted byproduct and killing innocents straight off become virtually redundant. As in the passage just quoted at the end of the last paragraph, one states the dilemma and seems then to fly by the seat of one's intuitionist pants.

Christian Pacifism

It would be unfortunate to leave the bishops on a sour note. On the one hand, they were reflecting the fact that serious Catholic philosophical and theological thought on war theory was very much alive and well. Engaging also with contemporary issues of importance. Georgetown University professor William V. O'Brien offered an in-depth analysis of "Desert Storm," the military action (1990–1991) against Iraq because of its invasion of Kuwait. Drawing on an impressive grasp both of Christian thinking and international law, O'Brien's conclusion was unambiguous: "The laws of war and the injunctions of humanitarian law were observed. Desert Storm was a just war" (O'Brien 1992, 823). On the other hand, commentators on the bishop's letter noted how strongly it is oriented toward the ideal of peace, not to mention the sympathy for pacifism—a sympathy that has not always been evident in the church's past history. Not that pacifism as such is necessarily a good thing, but that Christian clergy should take it seriously and perhaps even find points of merit is surely a good thing.

In this spirit, concluding this chapter, let us turn to the pacifists of our period. Again, perhaps not surprisingly, we find those who work from a purely Christian perspective, and those who while (sometimes) empathizing with the Christian aims, work from and try to offer reasons from a secular perspective. The Christian pacifist response to war is straightforward. Jesus said, most particularly in the Sermon on the Mount, that we should not meet violence with violence. Blessed are the peacemakers: for they shall be called the children of God. More generally, we should never use violence against our fellow humans. Remember:

> Ye have heard that it hath been said, An eye for an eye, and a tooth for a tooth:
>
> But I say unto you, That ye resist not evil: but whosoever shall smite thee on thy right cheek, turn to him the other also. (Matthew 5:38–39)

You cannot get more straightforward than this. Perhaps less straightforward than simple-minded. Is it not impractical for everyday living? What are we to do with Hitler? Czechoslovakia yesterday, Poland today, Britain and the rest of the world tomorrow. Here the pacifist reminds us that we are not alone, that God exists and is all-powerful, and that everything we do should be put in this context. Eschatology, end times, is all-important. Whatever Hitler may do is one thing. What we do is another. Our job is to focus on our relationship with God, as given to us through Christ. However much suffering our refusal to fight back may cause on this earth, it is but a prelude to the hoped-for eternal bliss with our creator. If we want to make sure that the second will come about, we have to be prepared to put up with the tribulations of the first. It is a religious appeal that is at stake here, not a doctrine that is merely a Christianized view of natural law, something that emerged from pagan thought.

Stanley Hauerwas (Methodist), one of the deeper thinkers in the second half of the twentieth century among those approaching war from a pacifist conviction, quoted John Howard Yoder (a Mennonite), another such pacifist thinker about war. Commenting on a statement by the American Methodist Bishops on war, showing just how far he was from a fellow Christian like Reinhold Niebuhr, Hauerwas wrote: "What the bishops fail to appreciate is that the peace that sustains Christian pacifism is an eschatological notion. Christian pacifism is not based on the assumption that Jesus has given us the means to achieve a warless world, but rather, as John Howard Yoder suggests, peace describes the hope, the goal of which in the light of which the pacifist acts, 'the character of his action, the ultimate divine certainly which lets his position make sense; it does not describe the external appearance or the observable results of his behavior.'" Continuing to quote Yoder. "This is what we mean by eschatology: a hope which, defying present frustrations, defines a present position in terms of the yet unseen goal which gives it meaning" (Hauerwas 1988, 436).

If Hauerwas was critical of the bishops of his own denomination, this was nothing to his scathing review of the statement of the bishops of another denomination, the Catholic bishops and "The Challenge of Peace." He went at once to the heart of the difficulties highlighted above. The bishops want peace, but then they offer rules for living in today's world, which they think may well demand war. It is this, rather than specific issues about deterrence, that lead to the paradox. "In order to develop a position capable of providing such guidance, the bishops, in spite of their analysis of the New Testament, turn to the just war theory. That theory, they argue, is built on the fundamental assumption that 'governments threatened by armed, unjust aggression must defend their people. This included defense by armed force if necessary as a last resort' (75). We therefore have a fundamental right of defense" (Hauerwas 1984, 402).

The trouble with this is that it is not grounded in Christianity but in the (pagan) idea of natural law. "For while their letter is addressed 'principally to the Catholic community, we want it to make a contribution to a wider public debate in our country on the dangers and dilemmas of the nuclear age.' Therefore, the ethical basis of The Challenge of Peace must be one that is not based on specifically Christian presuppositions" (401). Suppose we do want to put everything in a Christian eschatological framework. Does this mean that our whole lives are basically oriented toward something in the future? We are just putting in time down here? Hauerwas is at pains to say that this is not true. We are in some sense already living in the future. Amillennialism is the right theology. Relying again on Yoder: "The eschatology of the New Testament rests not in the conviction the kingdom has not fully come, but that it has. What is required is not a belief in some ideal amid the ambiguities of history, but rather a recognition that we have entered a period in which two ages overlap." Being pacifists on this earth, "points forward to the fullness of the kingdom of God, of which it is a foretaste" (420). If someone does a good act, it is not to buy the way into the kingdom of heaven, but to bring the kingdom of heaven down here on earth. Sophie Scholl of the White Rose Group going to her death for opposing Hitler—her acts and her bravery transcend the past, sinful age, and bring on the reality of what is to come. Amillennialism perhaps, but far from the optimism of the Darwinian.

Secular Pacifism

Turn finally for completeness to Robert L. Holmes, a nonbeliever who defends pacifism. In all he writes, one discerns the bitter aftertaste of Vietnam, beginning with the perceived failure of established churches to come out fully against that futile and soul-destroying war. He has little time for Christianity, for Augustine, or for just war theory. Above all, he finds just war theory in some important sense culpable. It makes smooth the way to sin. "While one cannot know what would have taken place had Christianity not charted for itself a new course after Augustine, the consequences of the course it did take holds increasingly grim prospects as they unfold in our modern world. It is as though belief in the pessimistic picture painted by Augustine has helped bring about the truth of that belief." Obviously, one doesn't want to blame Christianity for all of the ills of the world. "But had it clung to the nonviolence of the early Church, much of the war and destruction the Western World has known in the past fifteen hundred years could not have taken place" (Holmes 1989, 145). Holmes thinks the belief in our inherently sinful nature must take a lot of the blame. This is part of the Christian package. "But to locate the source of man's problems in human nature is, in effect, to reconcile oneself to their perpetuation. For then all that is left is to participate in evil and to confess one's own guilt in the process. You may then proceed in Christian love to slaughter your fellow men by the thousands. Wars then do become inevitable" (264).

For Holmes, just war theory breaks down generally, and especially in the face of nuclear weaponry. "To assume . . . that but for deterrence (by which I mean the system of deterrence—the possession, deployment, and willingness to use nuclear weapons by both sides) nuclear war might have occurred or might occur in the future is unintelligible. But for deterrence a nuclear war could not occur; it is the possession, deployment, and willingness to use those weapons that makes nuclear war possible" (258–259). In any case, war can never be justified, legitimately. You are always going to kill innocents, and that is wrong. Just war theory just takes you off on tangents. It steers you away from the fundamental issue: "modern war inevitably kills innocent persons. And this, I contend, makes modern war presumptively wrong" (211). Holmes adds that he has not shown definitively that no argument could be given to justify war; but he thinks none of the arguments so far put forward are convincing, and it is clear that Holmes does not really think that any good ones will ever be forthcoming.

Putting in Context

In an important way, it is good to end this discussion with Robert Holmes, a nonbeliever. More than Walzer, he is breaking away from the Christian perspective and tools of argument, and, in doing so and in showing what he finds so wrong, he supports our reading of the Christian position on war. Above all, it starts with our sinful nature. That is why war occurs, for all that we are made in the image of a good God. That belief is the bottom line for Christians across the spectrum, from those who are enthusiastic warmongers, through those who will fight if it is necessary, to those who are pacifists. The interesting emerging questions are now about the differences rather than the agreements. Hard-line Augustinian original sin does indeed confirm Holmes's worry and make war nigh inevitable, or at least (in Christian terms) empirically bound to happen. A realist like Niebuhr would say that not only is this the right way to view things, it is the morally sensible way to view things. Being Pollyanna is simply not helpful. The obvious next move for one embracing this position is to say that as Christians we must fight. It is though hardly less obvious that a pacifist (who accepted original sin) could simply acknowledge that war and violence will happen, but put things in an eschatological context. God expects us to turn the other cheek, no matter what happens. Although he does not accept Augustinian original sin, there are elements in this kind of thinking in Kant. He recognizes that his absolute prohibition on lying is not practical. But he argues that in the greater picture, God will put things right.

Not all Christians take this hard line. In the tradition of Kellogg and Fosdick, one does not read this in Paul Ramsey. Nor in the Christian pacifists we have discussed. We are sinners, but not necessarily hard-line Augustinian sinners. We can and must rise above ourselves, and—particularly in this age of nuclear weapons—find some way to move beyond war. Obviously, in a sense this is a move in the direction of a more progressivist world picture. "In the direction," perhaps, but it is certainly never going to be the world picture of someone like Julian Huxley. Whatever happens, whatever our free will, it all comes through the will and love of God. He will make possible any successes we might have. "God helps those who help themselves" is a saying recorded of Benjamin Franklin, not Moses or Jesus, but it is not an unfair summation of the way in which these Christians think.

10 | Darwinian Theory Comes of Age

THE YEAR 1959 WAS THE hundredth anniversary of the publication of the *Origin of Species* and the one hundred and fiftieth anniversary of the birth of its author, Charles Darwin. There were many celebrations, most notably at the University of Chicago, where, like candy at Halloween, honorary degrees were handed out to the founders of the synthesis (Smocovitis 1999). Julian Huxley responded by giving a humanist, evolutionary sermon in the campus chapel! The decades following, at the professional level, were to see an explosion of interest in Darwinian evolutionary theory, innovative theoretical ideas and empirical findings on a regular basis, and controversy within and without the field. Most significant in many respects was the coming of molecular biology, made possible by the Watson-Crick model of the DNA molecule. At first, traditional biologists regarded the molecular intrusions with suspicion, but soon it became clear that, properly used, the knowledge of life down at the particulate level could throw much light on hitherto-insoluble problems of evolutionary theory.

War continued to be a matter of much interest and concern, and as the professional science evolved and developed, so we see corresponding changes and discussions in the domain we are considering. Start with ethology.

Aggression

Social behavior was of great interest to Darwin, who never doubted that it is something shaped by natural selection. For a number of obvious reasons, it lagged rather behind other branches of the consilience. Most simply, behavior is hard to study. We all know how animals in zoos do not always behave like animals in the wild and, in any case, it is far less easy to document the mating pattern of a butterfly than to study its wing colors and patterns. Coming more from outside was the growth of the social sciences and their understandable push for control and power. Biologists were to be kept out! The assumption was that if you have seen one white rat you have seen them all. What need for more?

However, there were biologists who bucked the trend, notably the ethologists, that group of continental students of behavior of the 1930s, led by Konrad Lorenz and Niko Tinbergen. After the Second World War, conveniently forgetting past beliefs—later in life, these were acknowledged, very reluctantly, under pressure (Burkhardt 2005)—Lorenz showed a genius at writing for the public. Perhaps a belated voyage of self-discovery, he became much interested in conflict and violence, generally in animals and specifically in humans. His best-selling *On Aggression* (1966) went into the matter fully, looking at a major paradox. In animals, there are all sorts of mechanisms that stop violence from escalating within the species. There is conflict, true. At some point, however, something physiological kicks in and stops the progress to killing and death. Often, the loser or the weaker will show submissive behavior. This does not lead to instant death, but to a diminution of aggression and violence. "This is certainly the case in the dog, in which I have repeatedly seen that when the loser in a fight suddenly adopted the submissive attitude, and presented his unprotected neck, the winner performed the movement of shaking to death, in the air, close to the neck of the morally vanquished dog, but with closed mouth, that is, without biting" (Lorenz 1966, 133). Lorenz was a totally committed Darwinian, but Wallace-like in seeing selection as something that could work for the group over the individual. It is good for a species to order its members by strength. It does not pay it to kill off the weaker ones. "Thus in most cases the species-preserving function of the rival fight, selection of the stronger, is fulfilled without loss or even wounding of one of the individuals" (113).

Humans, alas, are the killer apes. We do not seem to have these inhibitions. We didn't need them, because the way in which we evolved meant that we simply did not have the weapons of aggression that sparked the need for inhibitions. We were biological wimps. Then that intelligence backfired.

> In human evolution, no inhibitory mechanisms preventing sudden manslaughter were necessary, because quick killing was impossible anyhow; the potential victim had plenty of opportunity to elicit the pity of the aggressor

> by submissive gestures and appeasing attitudes. No selection pressure arose in the prehistory of mankind to breed inhibitory mechanisms preventing the killing of conspecifics until, all of a sudden, the invention of artificial weapons upset the equilibrium of killing potential and social inhibitions. When it did, man's position was very nearly that of a dove which, by some unnatural trick of nature, has suddenly acquired the beak of a raven. (241)

All is not lost, however. That very same intelligence that led to weapons of destruction gave rise also to our moral sense and hope of control. Imagine a pre-human primate with a hand-axe, all sharpened up ready for use, meaning ready for killing, and no inhibitions. We would have wiped ourselves out almost immediately, were it not for the fact that along with the inventive abilities came moral or responsible abilities. Truly, both are facets of that same power of reason, of asking questions and finding answers. This opens up the possibility and necessity of bringing the good side of human nature to bear on the bad side. Then there is optimism for the future. "I believe in the power of human reason, as I believe in the power of natural selection. I believe that reason can and will exert a selection pressure in the right direction. I believe that this, in the not too distant future, will endow our descendants with the faculty of fulfilling the greatest and most beautiful of all commandments" (299).

Lorenz is running some changes on the usual picture. He sees war more as a later rather than earlier phenomenon. Notwithstanding, overall, he fits nicely. We have evolved with a dark side, but now thanks to our superior nature, there is hope of lasting peace.

Human Evolution

In the *Descent*, Darwin established beyond reasonable scientific doubt that we humans are part of the world of evolution (Ruse 2012). We descended from apes. Almost certainly not apes now extant, but apes nevertheless. There was debate about where our evolution took place. Darwin favored Africa, but Asia was at least if not more popular—no one much cared for the idea of Europeans coming from Africans—and when, almost at the end of the century, the first real "missing link" was unearthed, this seemed to be definitive. "Java man"—what we know now as *Homo erectus*—was found in the Far East and that seemed to be that. In the 1920s when Raymond Dart in South Africa began singing the importance of Taung Baby—what we now call *Australopithecus africanus*—no one was much impressed, not the least because likewise downplaying the importance of Africa were now incredible finds in England of the already-mentioned Piltdown Man, a being apparently with a human-type skull but an orang-type jaw (Falk 2012).

Finally, in the early 1950s, Piltdown was shown to be the fraud that it is—it looks like a human-orang mixture because it is a human-orang mixture! At the same time, paleoanthropologists started to open up the fossil record, especially in Central Africa. Praise above all must go to the indefatigable Leakey family, who turned up one pertinent fossil after another, although the prize find came in the early 1970s in Ethiopia. There the American scientist Donald Johanson and his associates uncovered Lucy, *Australopithecus afarensis*, now known to be somewhat more than three million years old, and remarkable not only for her chimpanzee-size brain—it was not a chimpanzee-like brain and seems to have been on the way to a brain like ours—but also for her distinctive upright stance (Johanson and Edey 1981). Showing that she really was a link, a hominin, Lucy could probably climb trees better than modern humans, and there is some evidence that her arms made her also better suited than we for tree life.

New theories, new findings, and yet when we turn our attention back to human war, it is remarkable—although perhaps by now we should not be surprised—how little the basic story changes. One of the most powerful and popular writers in the middle of the century was the playwright-turned-anthropologist Robert Ardrey. To his great credit, apart from having an ability to tell a good story in a way that is quite beyond the average scientist—average academic, in any field, for that matter—Ardrey got to know the scientists involved and what they thought and why. This was all fitted beautifully into the picture of man-the-killer-ape who now has the need and the opportunity to transcend his violent past. Raymond Dart—discoverer of *Australopithecus africanus*—was the chief authority and "The Predatory Transition from Ape to Man" (1953)—an article that "no regular scientific journal would touch"—was the key source of information. "What Dart put forward in his piece was the simple thesis that Man had emerged from the anthropoid background for one reason only: because he was a killer." We branched off from our "non-aggressive primate background." Because we had left the jungles and were stuck out on the plains, in full view, we had to learn to fight and given the evolution of bipedalism and the freeing of our hands, the making of weapons was almost predetermined. "A rock, a stick, a heavy bone—to our ancestral killer ape it meant the margin of survival. But the use of the weapon meant new and multiplying demands on the nervous system for the co-ordination of muscle and touch and sight. And so at last came the enlarged brain; so at last came man" (Ardrey 1961, 29).

We are killer apes. Interestingly, before we dash off with the conclusion that here we have, unfortunately, yet one more manifestation of the diseased thinking of (what the novelist Graham Greene memorably labeled) the "ugly American," if anything, Ardrey's enthusiasm for this kind of thinking stemmed from old-fashioned liberalism, rather than the contrary. He was one of those 1930s left-wingers who, like George Orwell, tempted by communism nevertheless saw its evil side. Hence, although he did suffer fallout from the McCarthy repressions against left-wing thinkers—to the extent that he pulled back from his career as a screenwriter in Hollywood and went to Africa in search of less troubled times and places—he hated the idea that human nature is infinitely malleable and thus open to indefinite change through societal forces. In short, he rejected the essence of the Marxist philosophy. But he was left wing, and did not want to leave it simply that we humans are and always will be upright apes, with blood streaming from our conquests over the weak and defenseless. Combining his social predilections with a firm belief in the powers of group selection, Ardrey (again in a rather traditional fashion) saw hope and light ahead for those who would seek and make the effort. At the individual level, being nice doesn't make much difference. At the group level, it can be all-important: "if the contest is between societies, then the member of a successful society must develop two sets of emotional responses: the many facets of friendship and co-operation reserved for members of his own society, and the many facets of hostility and enmity for members of the opposing society" (169).

In other words, we are almost literally two-faced and this offers hope to an end of war. "Human warfare comes about only when the defensive instinct of a determined territorial proprietor is challenged by the predatory compulsions of an equally determined territorial neighbour." Continuing that "the territorial drive brings about the conditions—not the motives—that give rise to war: the separation of men into groups, the alliance of men and territory, and the latent capacity for the enmity code to dominate the most civilized man in his relation to a hostile neighbour." However, there is, thank goodness, the other side to our nature, one "that may provide the foundations for a philosophical revolution." As we have our dark side, so also we have our light side. "The command to love is as deeply buried in our nature as the command to hate" (173). There is no guarantee that any of this will occur. There is hope: "civilization is a normal evolutionary development in our kind, and a product of natural selection." Without it, we would be back with the pre-human primate let loose with a sharpened hand-axe. Thus: "Civilization is a compensatory consequence of our killing imperative; the one could not exist without the other" (348). With the kind of sepulchral relish that seems often to overtake good liberals contemplating the future, Ardrey concludes with a gloomy description of human nature that would do credit to Thomas Hardy. "It is a jerry-built structure, and a more unattractive edifice could scarcely be imagined. Its greyness is appalling. Its walls are cracked and eggshell thin. Its foundations are shallow, its antiquity slight. No bands boom, no flags fly, no glamorous symbols invoke our nostalgic hearts." Shades of *Jude the Obscure*, and like poor Jude we're stuck with it. "Yet however humiliating the path may be, man beset by anarchy, banditry, chaos and extinction must at last resort turn to that chamber of horrors, human enlightenment. For he has nowhere else to turn" (352–353). At least, thank God—or Darwin—most of us, unlike Jude, do not have wives like Arabella.

Primates

The molecular revolution had wide ramifications. Almost paradoxically, one place where it was most strongly felt was in the study of the fossil record, paleontology or—as it has morphed with greater biological understanding—paleobiology (Sepkoski and Ruse 2009; Sepkoski 2012). Sewall Wright's theory of genetic drift, change that happens randomly beneath the effects of selection as it were, has had mixed fortunes when it comes to the genes producing regular physical features. However, down at the molecular level, much escapes the force of selection, and just drifts or changes, randomly, but at quantifiable rates (Kimura 1983). This "neutral theory of evolution"—because it says nothing threatening about the levels of selection that interest them, acceptable to and even welcomed by conventional Darwinians—has proven a powerful tool in judging rates of evolutionary change (Ayala 2009). Combined with earlier methods of calculating change thanks to radioactive decay, students of the fossil record now have huge amounts of information about exact dates when various momentous events occurred.

Using the neutral theory of evolution, checking on the data pertinent to our relationships with the higher primates—the chimpanzees (as well as the bonobos), the gorillas, and the orangutans—the unbelievable finding was that humans split from chimps about five million years ago. More shocking still is that we humans are more closely related to the chimps than they are to gorillas. (Recent studies push the dates back a bit, but not by that much.) All of this led to renewed interest in primate behavior, particularly the behavior of the great apes. Do we see in them significant clues to human behavior, and, with respect to our focus of interest, do we see significant clues to human warfare? The background assumption of many was very much that of Konrad Lorenz, namely that humans uniquely are killers and others have built-in safeguards against intra-specific violence. This was based on group selection—selection can benefit the group at the expense of the individual. Then things changed. Thanks to arguments by the American ichthyologist George Williams (1966)—later a major force behind evolutionary medicine—and to model building by the British, then-graduate student William Hamilton (in 1964), the "selfish-gene" perspective—one for one and to hell with the others—swept all before it. This was not because evolutionary biologists had all adopted the sociopolitico views of Ayn Rand. Rather it was because it was felt—as it is still felt by most Darwinian evolutionists—that group selection opens itself too readily to cheating. Given two organisms, similar except that one is selfishly serving just itself and the other altruistically giving (at its own expense) to the group, then selection suggests that the selfish organism will on average be more successful, even though in the long run the group would benefit. "Would benefit," but never will benefit, because of the preponderance of selfish individuals in the population.

With the demise of group selection, the picture might be expected to change. As indeed it did. By far the most famous of all primate studies—at least, of primate studies in the wild—were those of Jane Goodall (1986) in Africa on groups of chimpanzees. She followed these beasts around for literally weeks, months, years, and she found that unexpected patterns started to emerge. If you walk once down a street in Detroit, you are unlikely to see violence, let alone murder. If you walk down streets of Detroit for several years then the picture starts to change. It is the same with the chimpanzees. Goodall found that not only was there violence between chimpanzees, but that sometimes, in an almost systematic manner, violence took on the organized form that we associate with war. It is not just a matter of random encounters, but almost as though it were planned. So much for Lorenz's views about the group over the individual.

Goodall makes much of what she calls the process of "pseudospeciation," which seems to be a kind of cultural phenomenon where hitherto-integrated groups split into subgroups with cultural differences leading to hostilities between them. "Pseudospeciation in humans means, among other things, that the members of one group may not only see themselves as different from the members of another, but also behave in different ways to group and nongroup individuals.[1] In its extreme form pseudospeciation leads to the 'dehumanizing' of other groups, so that they may be regarded almost as members of a different species." She adds: "this process, along with the ability to use weapons for hurting or killing *at a distance,* frees group members from the inhibitions and social sanctions that operate within the group and enables acts that would not be tolerated within the group to be directed towards 'those others.' " Concluding, "this lack of inhibitions is a prime factor underlying the development of destructive warfare" (Goodall 1986, 532).

She goes on to say that these are precisely the sorts of things we find among the chimps, with nongroup members being "dechimpized." Goodall agrees that, with our megabrains, humans can think and plan in ways far beyond the other primates. This is a difference of degree and not of kind. "The chimpanzee, as a result of a unique combination of strong affiliative bonds between adult males on the one hand and an unusually hostile and aggressive attitude toward nongroup individuals on the other, has clearly reached a state where he stands at the very threshold of human achievement in destruction, cruelty, and planned intergroup conflict" (534). All the chimp needs is language, and he is not so far from that, and full-fledged war will be a reality. Man the killer ape is at the front of this picture—not because man is uniquely horrible, but because everybody is.

Culture Revives

Whether or not there are hopes of controlling it, a chorus of voices were singing of our violent nature. There was more. A few years after *On Aggression*, Lorenz's fellow ethologist and Austrian, Irenäus Eibl-Eibesfeldt, followed up in the same spirit. In his world, there is lots of intragroup aggression, tamed and controlled by ritualized behavior triggering hormonal effects that turn off violent feelings. "Man's intergroup aggression, however, generally aims at destruction. This is the result of cultural pseudospeciation, in the course of which human groups

[1] There is an ambiguity here in my use of the word "group." Group selection asserts cooperation (without hope or expectation of return) between non-relatives as in a whole species. Darwin (and others in his footsteps) allow such cooperation between relatives or presumed relatives, as in a tribe. Goodall is clearly thinking of cooperation in groups of this second sense, and denying cooperation in groups of the first sense.

mark themselves off from others by speech and usages, and describe themselves as human and others as not fully human" (Eibl-Eibesfeldt 1979, 240). Shades of the Nazis and the way that Goebbels portrays the attitude of good, clean, human Germans toward the evil, filthy, rodent Jews in *Jud Süss*. Fortunately, intelligence and reasoning start to kick in, and we set up ways of avoiding extreme bloodshed. Our biology and our culture must battle it out. There is hope. "The root of the universal desire for peace lies in this conflict between cultural and biological norms, which makes men want to bring their culture and biological norm filters into accord. Our conscience remains our hope, and based on this, a rationally guided evolution could lead to peace" (241).

People like Lorenz and Ardrey were writing just at a time when there were willing ears to listen. A perceptive recent writer notes the possible cause and effect. "In the 1960s and 1970s, as the Vietnam conflagration raged and the Cuban missile crisis menaced the Northern Hemisphere with nuclear Armageddon, Robert Ardrey (1908–80) and Konrad Lorenz (1903–89), and other dramatists and writers of popular books envisioned Dart's killer ape in every man, not just the ancient cave-dwellers" (Tuttle 2014, 5). It is worth pointing out that it was not just Dart's killer ape that was frightening people, but the constant fear that Augustine might be right about the taint of original sin. Dart himself started that article so influential on Ardrey with a quotation from the seventeenth-century Puritan divine, Richard Baxter: "Of all beasts the man-beast is the worst to others and himself the cruellest foe." Ardrey himself referred to us as "Cain's children" (Sussman 2013, 101). A fascinating riff on these themes is *The Planet of the Apes*, the 1968 movie, still spawning sequels. Given the iconic status of the final scene when the astronaut Taylor (Charlton Heston) discovers the battered Statue of Liberty and realizes that he has returned to his own nuclear-devastated planet many years after he left, one is hardly now spoiling things by revealing that the underlying theme of the movie is that humans are now primitives—compared to the intelligent and controlling apes—and that this is a function of our innate, warlike nature. The orangutan President of the Assembly, Dr. Zaius, tells all. "I have always known about man. From the evidence, I believe his wisdom must walk hand in hand with his idiocy. His emotions must rule his brain. He must be a warlike creature who gives battle for everything around him, even himself" (Ruse 2013, 161).

In the movie, this dark view of human nature—given wonderful faux theological backing—was final. Just as well, for it is hard to imagine sequels if Taylor had turned the corner to find a New York City traffic-free,

Central Park safe to walk at night, and the Metropolitan Opera capable of putting on a half-decent production of the *Ring*. In the Darwinian world came hopes that none of this conflict is permanent and we can see a way forward, out of the morass. Although there were the inevitable, ongoing debates about culture and biology. One with much to say was the Princeton-residing (he did not have a university affiliation) cultural anthropologist Ashley Montagu (1905–1999). A Jew—he was born Israel Ehrenberg—he had personal and political reasons to hate exclusively biological approaches to humankind. After the Second World War, he was involved in the UNESCO Statement on Race, where although it is allowed that there are biological differences between human beings, these are downplayed and the unity of our species is made the dominant theme. It was almost preordained that Montagu would take up the issue of violence and war, making the leitmotif the interplay and tension between the biological and the cultural—one way, von Bernhardi with his biological pessimism, and the other way William James with his cultural optimism. Given the appalling events of the first half of the century, it was vital for Montagu to show that violence is not biologically determined.

Montagu did not want to deny biology. "That human beings inherit genes which influence human behavior is a fact. It is also a fact that genes for basic forms of human behavior such as aggression, love, and altruism are the products of a long evolutionary history, and that in any serious examination of the nature of such forms of human behavior the evolutionary history of the species and its relations must be taken into account" (Montagu 1976, 308). This said, however, the unique thing about humankind is our intelligence and our abilities to transcend our biological heritage. Note however that this is not taken as a repudiation of Darwinism and a vote for Christianity. The very opposite in fact.

> Humans are neither naked apes nor fallen angels riven by that original sin, that great power of blackness, which Calvinistic commentators and their modern compeers have declared to actuate us. Neither are humans reducible to the category of animals for we are the human animals, a humanity which adds to being a dimension lacking in all other animals, creatures of immense and extraordinary educability, capable of being molded into virtually every and any desired shape and form. (315–316)

What this all means of course is that war is not inevitable or even contingently bound to happen. Explicitly, no Augustinian/Calvinist influences here. War is not predictably certain and we can rise above it. Why then have we been sinking into such violent conflict? Here Montagu brought up another of those mid-century cultural factors, so engaging liberal thinkers like Julian Huxley. Population numbers! Shades of Malthus, people obsessed over the growing human race. "It is not innate depravity that has brought humanity to this sorry pass, but the simple failure to understand that no living organism, and especially such a vulnerable creature as *Homo sapiens*, can go on multiplying uncontrollably without destroying its environment, and eventually itself" (322). We hear echoes of that haunting cry for *Lebensraum* and where it led us all. For all this, as a good Darwinian, Montagu ended optimistically. We can and will triumph. Progress is possible after all. "Let us always remember that humanity is not so much an inheritance as it is an achievement. Our true inheritance lies in our ability to make and shape ourselves, not the creatures but the creators of our destiny" (325).

Edward O. Wilson

It is wrong to privilege any one figure in this whole discussion, but in major respects, Robert Ardrey set the terms and the mood. There was a veritable tsunami of outpouring, popular and semi-popular, on the themes he was promoting. To name but three books: *The Naked Ape* (1967) by the ethologist Desmond Morris; *Human Aggression* by the psychiatrist Anthony Storr (1968); and *The Imperial Animal* by the sociologist Lionel Tiger and the anthropologist Robin Fox (1971). As we turn now to one of the true heavyweights in the Darwinian story—one who in respects set himself deliberately against Ardrey—first, grasp a nettle that you may think should have been grasped earlier. Ardrey was not a professional scientist. He was a playwright and a screenwriter and only then a reporter and amateur anthropologist. He got respect for what he did, and he himself defended much that he did on the grounds that not only had he worked hard at the subject—and no one could deny the truth of this—but that as a non-scientist this often gave him insights, a kind of intuitive empathy, that the narrow professional was lacking. With reason, he could say that professionals, to his credit and their discredit, were culpably slow off the mark when it came to human evolution. The fact remains that Ardrey was an amateur and—from our perspective—this is important. The theme of this book is that a major legacy of Darwin's thinking is the way that it took on the role of a secular religion. As such, it is not purely professional science but is in the public domain. Hence, you might argue that, far from being a handicap, the fact that Ardrey was not a professional scientist confirms the approach taken. In a way, this is true, but if Darwinism—meaning Darwinism as secular religion—gets too far from the professional world, then it loses its savor. What makes the story interesting and important is the intimate connection between Darwin's theory as professional science and the use made of this as a foundation for a secular religion. What matters is religion rooted in the science: not religion paying lip service to the science and wandering off on its own as it pleases.

This makes Harvard biologist Edward O. Wilson crucial and central to our story. On the one hand, no one could deny him his status as a professional Darwinian evolutionist—indeed, he has claim to being the world's leading, professional, Darwinian evolutionist. He has a huge amount of work—theoretical and empirical—to his credit, from analyses of island biogeography, through studies of insect methods of chemical communication, to synthesizing of the whole area of social behavior considered from an evolutionary perspective (sociobiology), to detailed studies of ants and their caste systems, and much more (Wilson 1994). On the other hand, however, he has long been a man who wants to take his work into the public domain, today being much concerned with biodiversity and the devastation of natural resources, especially in South America. Wilson commands our immediate attention because, as we saw in chapter 2, for all that he was a professional scientist, indeed precisely because he was a professional scientist, he felt able to state unambiguously that he wanted to use Darwin's ideas as the basis of a new humanistic religion. He made this declaration, just after *Sociobiology: The New Synthesis*, in the Pulitzer Prize-winning work, *On Human Nature* (1978). He stated openly in the preface that this book was not as such an exercise of science: rather, a book about science, where he applied his thinking to our own species. Scientific materialism wins every time over religion. It is Amazon against your local, independent bookstore. The bookstore long had a full and functioning life; now, in this Internet age, it is obsolete. Its day is past. Move on, not reluctantly but with hope and optimism. "If this interpretation is correct, the final decisive edge enjoyed by scientific naturalism will come from its capacity to explain traditional religion, its chief competition, as a wholly material phenomenon. Theology is not likely to survive as an independent intellectual discipline" (Wilson 1978, 192). As God's sovereignty and mercy is the underlying metaphysic of Christianity, so equally human-driven advance is the underlying metaphysic of Wilson's Darwinism. "Progress, then, is a property of the evolution of life as a whole by almost any conceivable intuitive standard, including the acquisition of goals and intentions in the behavior of animals" (Wilson 1992, 187).

There is a strong odor of Spencerian thought in Wilson's views on science, particularly in his thinking about progressive rise in evolution. With this, going back to Spencer himself, one finds an almost casual belief that progress is bound to happen and that there is no big need to offer causes. It is just the way of things. None of this should be a great surprise because Wilson's intellectual grandfather—the supervisor of his supervisor—was W. M. Wheeler, a huge Spencer enthusiast (Gibson 2013). It should be no great surprise either, given Wheeler's predilections, to find that Wilson jumps right into discussion of social issues. Prominent among these is human conflict. Wilson certainly sees war as a natural thing. Part of our biological nature. "Intertribal aggression, escalating in some cultures to primitive warfare, is common enough to be regarded as a general characteristic of hunter-gatherer social behavior" (Wilson 1978, 82). We are by nature naturally aggressive. "Only by redefining the words 'innateness' and 'aggression' to the point of uselessness might we correctly say that human aggressiveness is not innate" (99). Where Wilson sets out to show his professional superiority over someone like Ardrey is through his knowledge and use of contemporary discussion about the level at which natural selection operates (Weidman 2011). Wilson unlike Ardrey knows all about the reasons for individual selection over group selection, and in the light of this offers three possible hypotheses. First, warfare is biologically irrelevant. "Cultural traditions of warfare in primitive societies evolved independently of the ability of human beings to survive and reproduce." Second, warfare is selected because of its value to the individual. "Cultural traditions of primitive warfare evolved by selective retention of traits that improve the inclusive genetic fitness of human beings." Third, warfare benefits the group. "Cultural traditions of primitive warfare evolved by a process of group selection that favored the self-sacrificing tendencies of some warriors" (110). By reference to specific primitive groups, Wilson has no hesitation in opting for hypothesis two. War is an adaptation benefiting the individual. At least it was in the case of primitive societies.

Now of course things are different. The classic Darwinian picture emerges. It is far from obvious that war is such a very good thing. "Although the evidence suggests that the biological nature of humankind launched the evolution of organized aggression and roughly directed its early history across many societies, the eventual outcome of the evolution will be determined by cultural processes brought increasingly under the control of rational thought." Sagely, using language that gave him street cred as a professional, Wilson added: "The practice of war is a straightforward example of a hypertrophied biological predisposition" (116). In other words, we must use reason to get around the uncomfortable and unwanted aspects of our animal nature. "We can only work our way around them. To let them rest latent and unsummoned, we must consciously undertake those difficult and rarely traveled pathways in psychological development that lead to mastery over and reduction of the profound human tendency to learn violence" (119). Modern dress, but the same story. Aggression is natural. This leads to war. It had to be a good thing in that it was part of the progressive process leading to us. Now, war is not a good thing, and it can and must be overcome by pressing the right cultural buttons.

The story does not end there. Notoriously, and almost humorously, Wilson has changed his thinking about the way in which selection operates (Wilson and Wilson 2007). To the amazement and the chagrin of his fellow Darwinians, Wilson began to enthuse about the possibilities of group selection and now is a full-blown advocate of so-called multilevel theory, meaning that he sees selection sometimes working for the individual and at other times working for the group (against the immediate interests of individuals). One should say that the move to this way of thinking did not bring forth abject apologies to the ethologists and fellow travelers, nor does it seem to have much affected Wilson's thinking on war. He continues to see biological virtues in aggression, particularly in the context of fighting for territory. "Since the control of limiting resources has been a matter of life and death through millennia of evolutionary time, territorial aggression is widespread and reaction to it often murderous" (Wilson 1998, 171). Benefit for the group is benefit for the individual, and conversely. Fortunately, despite this biological heritage, change is possible. It won't be easy: "war arises from both genes and culture and can best be avoided by a thorough understanding of the manner in which these two modes of heredity interact within different historical contexts."

Frans de Waal

What Jane Goodall did for studies of chimpanzees in the wild, the Dutch primatologist Frans de Waal did for studies of chimpanzees in captivity. There is an important difference, for whereas the message we take from Goodall is about the naturalness of ape violence, the message we take from de Waal is about the naturalness of ape cooperation. His most famous study, recorded in *Chimpanzee Politics* (1982), is of a captive troop of chimpanzees (about twenty-five). They were kept in a two-acre compound in the zoo, at Arnhem, in Holland. Conducted in the late 1970s, de Waal records in great detail and with much brio the interactions within the troop. The main story centers on the struggle for dominance between the three mature males, Yeroen, Luit, and Nikkie. At first, Yeroen was the alpha male. This can have huge reproductive payoffs so from a biological viewpoint it is a status well worth having—the only thing worth having! Then, through fighting and other forms of aggression, Luit made increasing moves toward dominance. In the end, he was successful and Yeroen was pushed aside. The story continues. Now, it was Nikkie's turn to move to the alpha role, which he did but only through an alliance with Yeroen, who clearly saw some payoff, if limited.

Complementing this struggle, there were the females who were far from passive bystanders. If the males were struggling on a ladder, then the females were enmeshed in a network. This was not just a network between females, but also bringing in the males. On the one hand, the males wanted respect and submission from the females. On the other hand, the males needed the support of the females as they tried to move up the dominance hierarchy. Some females, Mama particularly, were dominant over others. These were more cherished by the males than others. This all had payoffs, for instance, for the children of dominant females who got added status. Also, though, the more dominant females could act as peacemakers. For instance, if the children of two lower-caste females started quarreling, this could be bad all around, most worrisome being that the large, adult males might get annoyed and start (literally) throwing their weight around. The lower-caste females might be unable to solve the situation themselves, but could get a higher-caste female to intervene and stop the squabbles.

De Waal is open both in *Chimpanzee Politics* and in later writings that his basic position is that chimpanzees practice restraint and indulge in politicking to keep the peace, thus avoiding all-out conflict. Of Luit's challenge to Yeroen, de Waal writes: "We may wonder why the two rivals did not put an end to their conflict by fighting it out, once and for all. The answer is simple: because physical strength is only one factor and almost certainly not the critical one in determining dominance relationships" (87). There is more than this. There is also the need to keep a sense of community, even between rivals. War with outsiders is an ever-constant threat. "[T]he chimpanzee male psyche, shaped by millions of years of war in the natural habitat, is one of both competition and compromise. Whatever the level of competition among them, males count on each other against the outside" (105).

This is all starting to sound very familiar. War was part of our evolutionary past, but thanks to war we are on our way up to biological adaptations of restraint and ultimately morality. "Obviously the most potent force to bring out a sense of community is enmity toward outsiders. It forces unity among elements that are normally at odds" (54). De Waal stresses repeatedly that he is on the side of biological underpinnings to the thought and behavior of restraint, against people like Thomas Henry Huxley whom he sees, with reason, as being on the cultural side, thinking only this can rein in the savage ape within. He also stresses repeatedly that what we see in the chimpanzee can be transferred virtually directly to humankind. "In our own species, nothing is more obvious than that we band together against adversaries. In the course of human evolution, out-group hostility enhanced in-group solidarity to the point that morality emerged." Of course, we humans climbed a little (or a lot) higher. "Instead of merely ameliorating relations around us, as apes do, we have explicit teachings about the value of the community and the precedence it takes, or ought to take, over individual interests" (54). We are primed for the nigh-inevitable conclusion.

> And so, the profound irony is that our noblest achievement—morality—has evolutionary ties to our basest behavior—warfare. The sense of community required by the former was provided by the latter. When we passed the tipping point between individual interests and shared interests, we ratcheted up the social pressure to make sure everyone contributed to the common good. (55)

If this is not value impregnated—"noblest," "basest," "ratcheted up"—then it is hard to know what would be. Confirming that de Waal is offering us something that goes beyond objective, Popperian science is the revealingly inauthentic nature of *Chimpanzee Politics*. As de Waal admits in the twenty-fifth anniversary reissue of the book, even as it was being completed, he knew that there was more to the story (de Waal 2007). A far, darker and grimmer more to the story than one might have guessed. In the summer of 1980, Nikkie—then alpha male—began to lose control. He had made the mistake of cutting off Yeroen's access to fertile (estrus) females. Yeroen switched his allegiance to Luit who became again the alpha male. This lasted only ten weeks because one night—the three males were as usual sleeping together, separated off from the females—Yeroen and Nikkie apparently systematically (they themselves suffered little harm) turned on Luit and inflicted such damage that he died the next day from loss of blood. Among the damage was castration, for his testicles were found on the cage floor.

Although this was public knowledge, de Waal kept it out of the book. Why? "I decided against recounting it in the first edition of *Chimpanzee Politics* so as to avoid ending the book on a dark note" (de Waal 2007, 211). Well, yes, but . . . The impression we are left with is that compromise and conciliation work. Apparently, they do not, at least not all of the time, and not all of the crucial time. Had we but known about this, we might well still have agreed about the importance of evolved sociality. We might also have tempered our acceptance of the impression that although war is a bad thing, evolution has the solution well in hand. I do not particularly condemn de Waal on scholarly grounds. He adds that "at the time I was not yet emotionally ready to analyze this shocking event." We can all well understand that. He did discuss it in detail in a later book, *Peacemaking among Primates*, by which time he was ready to fit the story into the overall scenario, namely that we see the vital need of peacemaking adaptations. I am saying that the episode reinforces the claim that we have more here than straight science and something more akin to religion, or a secular religious approach. I have just written a book showing the goodness of God and then my wife dies of cancer. I am not about to bring it into this book—"I was not yet emotionally ready to analyze this shocking event"—but a year or two later I wrote another book where I showed my wife's death is all part of seeing this God-created world as the "vale of soul making" (the poet Keats's answer to the problem of evil).

Sociobiology and the Bomb

Do we find Darwinian concerns about war overlapping with those of the Christians? We have seen already suggestions that this was almost certainly so, particularly shared worries as the Cold War persisted and as Vietnam heated up. This continues, although expectedly there are special Darwinian themes. Starting in the mid-1970s, there was a major row over sociobiology, with Wilson pitted against hard-line leftists for promoting what they took to be (that already-mentioned) "genetically deterministic" view of behavior, particularly human behavior (Ruse 1979b; Segerstrale 1986). Leading this charge were two of Wilson's Harvard colleagues—they were even in the same biology department—the Marxist biologists Stephen Jay Gould and Richard Lewontin (1979). This led to an internal conflict of some irony. Eager as Gould was to push a left-wing line, he was as eager to promote the professionalism of his own particular area of evolutionary biology, paleontology—something traditionally rather put down by other evolutionists, particularly those working on genetics. Just then starting his dizzying climb as one of the great science popularizers, in his first collection of essays, *Ever Since Darwin*, Gould penned a piece that was equally a Wilson-like critique of the likes of Robert Ardrey to presume to write authoritatively on evolutionary biology, as it was a critique of the whole approach of Ardrey (and, by implication, Wilson). In "The Non-Science of Human Nature," we learn that Ardrey's "pop ethology" is based on "highly disputable" supposed evidence, nothing of which holds up. Showing that *Planet of the Apes* was not the only movie that picked up on the killer-ape theme, Gould noted gloomily that "Ardrey's dubious theory is a prominent theme in Stanley Kubrick's film *2001*," where the opening story is all about our ancestors learning to use weapons and thus to win in war.

The Gould-Lewontin attack on sociobiology extended out against others, with the University of Washington biologist David Barash picked out in a bright, critical spotlight. They charged that he was doing truly dreadful science, finding all kinds of fictitious Darwinian reasons to explain social behavior. Paradoxically, hard-line Darwinian evolutionist of social behavior though Barash most certainly is, for all that Gould and Lewontin were berating him as part of their general campaign against sociobiology as something merely a cover for pseudoscientific rightwing sentiments, Barash is as reliably left-wing as any good old comrade might wish. Just turn to the book on nuclear weapons, *The Caveman and the Bomb* (1985), he authored shortly thereafter, written with his psychiatrist wife, Judith Eve Lipton. A plea to stop the madness of nuclear weaponry, Barash and his wife made it clear from the beginning that their argument is rooted in science but is more than science. "In this book, we have purposely avoided a strictly scientific approach, because we have become convinced that science has significant scientific limitations, particularly when confronting social problems." However, science is crucially in the background. "On the other hand, we base our argument largely on a scientific assumption, that human beings evolved with the rest of life on earth, by an astounding but unmagical assumption—resembling the Tao in its passivity and universality—called natural selection" (Barash and Lipton 1985, x). An argument that at once starts to have familiar echoes. "We further argue that our current nuclear predicament derives largely from a fundamental gulf that characterizes much of the human situation: the disparity between the structure of the human mind and the realities of our current environment." Fortunately: "Give man and womankind a sufficiently broad and varied framework, as well as time to think it through, and we believe that the potential exists for us to transcend both biological and cultural traditions, and to become truly "sapient" at last" (xi).

Barash and Lipton suppose what they call the "Neanderthal mentality," the mind and character of the human of evolutionary history. This is not an entirely happy conceit for all that people in Europe have picked up about 5 percent Neanderthal genes. Neanderthals are not our ancestors. One needs a plea of poetic license. The point is that these supposed Neanderthals evolved with all sorts of biologically engrained adaptations—Barash is not a sociobiologist for nothing—that were once useful but are dangerous today. Especially dangerous today as we now face the magnitudes of more destructive power of nuclear weapons. No doubt continuing to incur the ire of Gould and Lewontin, Barash and Lipton saw some of the adaptations as occurring within our species, for instance—thanks to sexual selection—the bigger size and power and aggressiveness of males. These features are not necessarily going to be that helpful when faced with the delicate deliberations required now given the Bomb. The more submissive nature of females does not help either. We need them to step up to the plate more and get involved in the discussion—"the world will be a safer place if and when women become more in tune with the life-affirming possibilities in their nurturant, biological possibilities" (18).

Other Neanderthal legacies are more species-wide and were once very important and cherished and promoted by natural selection. These today pose grave threats because they are adaptations forged in the fire of war—something of which our authors ask rhetorically: "Does this mean that primitive war was *good*?" Not at all, they reply strongly. They do allow that it "was good, however, for certain genes under certain circumstances" (27). You might in fairness note that this is a comparative judgment, but we have seen often—too often—how the comparative slides easily into the absolute. Among these adaptations brought on in response to war, they start with group identification. Note that this is considered from a selfish-gene perspective. It is better for an individual to be part of the group than otherwise—"there has always been security in numbers, and most of the time, safety as well" (62). The trouble is that this is all a bit like a craving for sweet things. Good in the Pleistocene. Terrible today. Except, I suppose, if you are a dentist. "Similarly, the modern-day Neanderthal satisfies his craving for group identification despite the fact that in an era of nuclear-armed nation states such identification may be not only useless, but downright dangerous" (65).

Paranoia, apparently, is something with its adaptive virtues. Don't trust anyone. They may be out to get you. Good attitude to have in the past apparently, but not necessarily today. "Our thesis is that because of our biological ancestry, given certain historical and cultural factors—notably conditions of stress or trauma—the policies and practices of nations may come to be dominated by paranoid attitudes, which in turn result in self-perpetuating cycles of aggression and war, more or less as we have already seen for individuals" (102). And so it continues through the sorry list. Deterrence was a good thing, but today is simply not genuinely possible. You have to be willing to use the Bomb and that in the end defeats the very purpose of what you about. Almost expectedly in discussions such as these, Barash and Lipton manage to get in a swipe at Christianity. So much of the thinking behind deterrence is predicated on the belief that humans respond only to threats and fear and never to reason. Predicated on the belief that humans are inherently vile and evil—original sin. This in itself is inherently dangerous. "So long as we see human nature as ravening, evil, and dangerous, a rogue elephant straining to trample our gardens if we relax our guard, we will keep spraying the perfume and building the bombs" (179).

Can nothing be done? Barash and Lipton find comfort and support in toilet training. It doesn't come naturally and other animals have huge problems with it. It can be done, and it is clearly better to have such training, for all that in the good old days you could relieve yourself wherever you found yourself in the jungle. Likewise with inclinations to war. Three factors that are pertinent. First, intelligence. One thing that history tells us is that war is not inevitable, but that if you are engaged in an arms race it is a lot more likely. You have all of those lovely weapons. Let's use them. So, as a counter, let us decide reflectively to stop the nuclear arms race. It has at times been done in part. Now let us make this universal. "If we unfetter our rationality from its Neanderthal taskmaster, and search diligently and honestly for solutions, we are led to some enormously hopeful conclusions—namely that the arms race can be halted and reversed, and that this will leave us much safer (and richer, incidentally) than we are today" (209). Then there is the appeal to feeling. Let's get natural selection involved here. We want to protect our relatives and friends—"insofar as the Neanderthal mentality has a genetic basis, each human being should be inclined to protect his or her relatives and future generations" (226). Ban the Bomb!

Finally, perhaps a little incongruously given what has gone before, there is the appeal to ethics and religion. Ethics seems obvious. Destroying the world could never be morally good. In the case of religion, it seems it is more a question of dismantling what we have than turning to it for insight. Particularly heinous is the Augustinian doctrine of a just war. This is just an excuse for people to go out and kill other people. In any case, most of the crucial insights—about not harming others—can be achieved without religion. More than this, religion has an unhealthy attitude toward death, almost thinking it to be a good thing. (Shades of Archbishop Fisher!) But still, properly understood, religion can play a positive role in the process. We must succeed. We can succeed. "Maybe in the long run, we shall all laugh together, as through our negation of the Neanderthal mentality we arrive at a new affirmation, a higher level of life, its most exalted accomplishment" (267). At this stage of our discussion, it is hardly still necessary to point to the optimistic belief in progress behind this concluding declaration.

Pacifism

In important respects, we have already encountered much Darwinian argumentation pointing, if not to outright pacifism, to a greater hope of peace than we find among the Christians, at least among the Augustinian Christians. That is a key part of the Darwinian approach to war, namely that we can transcend our biological past and point toward a better future. We have taken note of the fact that the Chicago Quaker biologist W. C. Allee was not coming from the same theological corner, he did not subscribe to Augustinian original sin, and this gave him a somewhat different take on the war business. He did not see war as part of our biological nature and thought that its elimination is a matter of cultural adjustment. There is no undue optimism about culture overcoming biology because there is no biology to overcome in the first place. A significant factor in Allee's Quakerism was the way in which it inclined him to holistic thinking about biology generally and selection specifically. He had a broad cultural effect. Through a close biologist friend—a former student at Chicago—the novelist John Steinbeck was fully aware of the kind of thinking that was so important for people like Allee—in terms of the group rather than the individual—and it is the philosophy underlying his great novel of the Depression, *Grapes of Wrath* (Ruse 2017a).

Allee took things further than others, but he was not a total outlier. His fellow biologist, the Englishman W. H. Thorpe, also a member of the Religious Society of Friends (Quakers), sang a similar song. This is a happy metaphor, for Thorpe's deserved claim to fame was how he recorded bird songs and then analyzed the tapes to yield quantifiable ways of treating what had hitherto been a very subjective area of study. In approaching someone like Thorpe—I speak with the authority of a coreligionist at the pertinent time—one needs to take care in thinking about what Quakerism would have meant to him, especially around the middle of the last century. Quakerism has always been very inviting of science, partly on theological grounds but also on cultural grounds. Friends were not allowed to join the military or the church, and so many went (very successfully) into business, and that meant science and technology. Combine this with the central place of mysticism, always a big thing for Quakers, given that theirs is a religion that relies heavily on personal experience of the divine—"the inner light." Almost inevitably, Thorpe saw an intimate connection between Christianity and evolutionary thinking—"*among world religions only Christianity can meet the demands of a scientific world view cognizant of evolutionary biology and only Christianity can serve a mankind fully conscious of its past and of the evolutionary possibilities of its future*" (Thorpe 1962, 119, italics in original).

Always, there is this deep, intense spirituality to the Quaker vision of the world. This is not the world of *res extensa*. We cannot get away from the divine, as creator and as loving sustainer. "To my mind, then, any rational system of belief involves the conviction that the creative and sustaining spirit of God may be everywhere present and active; indeed, I believe that all aspects of the universe, all kinds of experience, may be sacramental in the true meaning of the term" (Thorpe 1961, 58). This is why Quakers do not take holy communion. All of experience is a sacrament. In short, don't take the lack of ecclesiastic trimmings to mean that someone like Thorpe was really a Unitarian, whose idea of the Eucharist is singing Kumbaya and tying yellow bows on environmentally threatened oak trees. Thorpe was a deeply committed Christian (Thorpe 1968). So really, it is almost meaningless to ask if he believed in progress or Providence. In the end, they are all the same and in God's hands. Embracing evolution with this enthusiasm, no wonder that Thorpe felt able to take up ideas anathema to Augustinian Christians. Quakers have never had a hang-up about Augustinian original sin. That is why they are totally indifferent to the putative virginity of Mary. Jesus could no more be born of sin, even if Joseph was his biological father, than any of the rest of us. Progress to humankind was a central part of Thorpe's world vision, although note how, taking up ideas no less anathema to Darwinian nonbelievers, he puts this in the context of enthusiasm for the neo-Bergsonian thinking of the Jesuit priest Teilhard de Chardin.

> To my mind, the only conceivable answer to the admitted evil and suffering of the world is in fact the answer that Christianity has always given, that creation groans and travails until now. That it is an integral part of the process only acceptable to the mind of man in so far as he begins to see its place in a scheme so stupendous that even the uttermost depths of evil are subject to redemptive action. (Thorpe 1966, 58)

One almost expects a man like this to have trouble with war. Quite apart from the extra-scientific injunctions of Jesus against violence, how could it occur in the science-inspired, natural theological universe of Thorpe's God? Concerning what he called the "pornography of violence," Thorpe wrote, "while I am sure that the more we can do to control and prevent the atmosphere of violent propaganda the better; this is perhaps only one part and perhaps the least important part of the very profound fact that *Homo sapiens* seems to been infected almost from the beginning not with an instinctive drive to violence, but with a nature that makes it fatally easy for his socio-control methods to slip over into violent forms" (Thorpe 1974, 267–268). There is biology there, but it is not the biology of violence. It is the biology of being overly susceptible to bad ideas. Although he and Ashley Montagu are on the same side, Thorpe is not very taken with Montagu's thought that the root problem is overpopulation. Perhaps relatedly, however, Thorpe does suspect that pollution might be part of the problem. One imagines that, for a bird biologist, pollution is a more pressing threat than population. In the end though, Thorpe agrees with Konrad Lorenz, that grave though the situation may be, "the ultimate power of human common sense can hardly be over-estimated. If we have that common sense expressing itself in social and natural selection then the grounds for hope are much firmer than perhaps we think" (Thorpe 1966, 129).

Sex and War

Darwin thought it was all about sex—well, at least a lot about sex. As does Malcolm Potts (working with a ghostwriter), author of *Sex and War: How Biology Explains Warfare and Terrorism and Offers a Path to a Safer World* (2010). Resisting the temptation, on the basis of the subtitle, to declare victory for the thesis of this book and to head out for a beer, let us take a quick look at what he has to say, as a way of looking at something very much in the footsteps of the founder and to wrap up the chapter. Without irony or flattery, Potts comes across as a man of very great moral worth, for the book opens when he was working as a doctor in what is now Bangladesh, at the end of its war of independence from Pakistan. Obviously coloring his whole approach to war, Potts was treating girls who had been raped by Pakistani soldiers and he was pondering what it all meant. The rapes were horrific, not just because of the violations and consequences, pregnancies, but also because of the social effects and ostracisms of women who had been thus violated. Yet, the paradox is that these rapists were not your classic, lust-driven perverts, hiding in the bushes in parks late at night, waiting for unsuspecting victims. In their usual lives, they were conventional, law-abiding members of society. Then they found themselves in a situation where rape was possible and encouraged. Convention changed and they changed. Something, presumably something biological, kicked in.

It is clear that this something biological has to be something biological fashioned by natural selection and the reason is easy to find. Men in power, men who can get away with it, copulate and reproduce and that is what it is all about. Mention is made of Genghis Khan and his achievements in that direction. Elaborating, we learn that aggression—and Potts sees this as very much a male phenomenon—is something most efficiently channeled through group action. "Human warfare and terrorism require a special sort of violence in men, which we will call team aggression" (Potts and Hayden 2010, 11). Much of the book, expectedly, is about the evolution of groups or troops, and as expectedly the sorts of studies as reported by Frans de Waal get much prominence. As do the ways of soldiers in the last century's conflicts. Particularly striking were American bombing crews, who faced near certain death, but who would never have let down their buddies. How come? "Could it have been the primate predisposition for small groups of men to show enormous loyalty to each other, great courage in the face of death, and a lack of empathy in attacking an outgroup?" (64)

So battles begin. Women spur the driving force. Continuing to stay on much-traveled ground, we move from the chimps to the native people of South America. The top-ranked Yanomamö warriors, the most brave and efficient, get the goodies. It is they who get the most wives—over twice as many as normal men—and it is they who have the most kids—at least three times as many as regular guys. It continues. With the coming of technology and the like, everything gets jacked up into full-scale war, where the original biological drives get buried in the sociopolitical needs and desires of modern societies. Leading to the mess we find ourselves in today, and a familiar warning from our author. "If we are to survive our own deep impulses, and perhaps take the next step on the ongoing march of human progress, then the urge for self-defense and revenge must be tempered by cool, objective analysis of risk; measured, productive political responses; and military action only when absolutely necessary, appropriate, and effective" (174).

As women are the problem, so they seem to be the solution. Men are the aggressors. Women are the peacemakers. We must bring women far more to the front in society, empowering them through such things as education and proper medical care—"meeting the desire" of "family planning and safe abortions." Thus, we can deflect men from their innate dispositions. "This is about the most profound insight to come from taking an evolutionary perspective on war: empowering women reduces the risk of violent conflict. Far from being a politically correct notion of feminist philosophy, women's role in reducing the risk of war is borne out by rigorous study and historical experience" (14). This is to be backed by the fruits of progress. "If we don't destroy one another with war first, then a thoughtful application of scientific knowledge could solve most of our other problems, serious as they are" (382). I don't really think Potts is so pessimistic as to think we will destroy one another. He forecasts that we will tame our "Stone Age behaviors" (383).

All very familiar although nicely illustrating how religion or religious-type perspectives take on the cultural coloring or issues of the day. One thinks for instance of how Christians wrestle and continue to wrestle with such issues as homosexuality. Moreover, even though there are generally shared concerns, there isn't one set solution. An evangelical and a liberal Anglican can differ fundamentally on such issues, but they are still both Christians. Potts worries that his approach manifests "a politically correct notion of feminist philosophy," with the implication that it isn't really that at all. But of course it is! What Potts misses is that it might also be true or at least the best strategy. My point here though is to show how Potts is clearly sensitive to his culture—his values are showing through—and to say that this is a mark of religion, as also is the fact that a powerful religion can encompass quite a level of disagreement within its boundaries. Charles Darwin likewise thought that men competing for women is important. Remember: "With savages, for instance the Australians, the women are the constant cause of war both between members of the same tribe and between distinct tribes" (561). Darwin's solution, inasmuch as he saw war as a problem, was not to empower women. He would have thought that absolutely madness. As would his wife Emma, who ruled the Darwin family as a benevolent despot.

A Thesis Confirmed

With this last author, we are getting to the stage where the main points are confirmed with almost monotonous regularity. We start in science but we end drenched in values. Putting on one side the significant and ongoing minority position about war not being innate—this position will not be forgotten—the story is simple. War is part of our biological past. It led to humans as they are today. Already we have seen the influence of an article, in *Science*, by the evolutionary anthropologist Napoleon Chagnon. This is about those tribes, in all about 35,000, who live in the Brazilian rain-forest. They are folk who take killing each other seriously. "Studies of the Yanomamö Indians of Amazonas during the past 23 years show that 44 percent of males estimated to be 25 or older have participated in the killing of someone, that approximately 30 percent of adult male deaths are due to violence, and that nearly 70 percent of all adults over an estimated 40 years of age have lost a close genetic relative due to violence." And it is all fueled by and for natural selection! The more successful you are at killing, the more wives you tend to have, and hence the more children. We have a beautiful example of what another evolutionist calls "the natural good of evolution itself" (Domning 2001, 109). This all said, war today is no longer biologically or culturally desirable, or even allowable. We can and must do something about this.

The literature is large and the sampling of this chapter select. But not unrepresentative. Picking almost at random a much-regarded work on the topic, co-authored by Harvard biological anthropologist Richard Wrangham, the familiar themes and tropes appear. Males are aggressive and natural selection made them that way. It is not a good state of affairs today. Fortunately, there is hope. "Male demonism is not inevitable. Its expression has evolved in other animals, it varies across human societies, and it has changed in history" (Wrangham and Peterson 1996, 251). It isn't going to be easy, but look ahead with confidence. "We can have no idea how far the wave of history may sweep us from our rougher past." Familiar problem, familiar solution. Sometimes, the confirmations are almost spookily on target, to use a military metaphor. A recent, acclaimed book by Stanford classicist and historian Ian Morris has the central thesis that war is a good thing! Appearances to the contrary, "war has been good for something: over the long run, it has made humanity safer and richer" (Morris 2014, 7). The author even has "progress" in his title! There are four pertinent points. First, fighting has led to larger societies, better organized than before, thus reducing the risk of violent death. This holds true even of the last dreadful century. "If you were lucky enough to be born in the industrialized twentieth century, you were on average ten times *less* likely to die violently (or from violence's consequences) than if you were born in a Stone Age society" (8). Second, war seems to be the only way—certainly, the only way we have found—to make for bigger and better societies. People just don't do good things for other people unless they are forced to and the best of all forces is "defeat in war or fear that defeat is imminent" (9). Third, larger societies make us richer. "By creating larger societies, stronger governments, and greater security, war has enriched the world." And fourth—this sounds familiar!—societies have now developed to such a stage that war is out of date. We don't need it anymore and it is downright dangerous and threatening. "For millennia, war (over the long run) has created peace, and destruction has created wealth, but in our own age humanity has gotten so good at fighting—our weapons so destructive, our organizations so efficient—that war is beginning to make further war of this kind impossible" (9). You will be expectedly relieved to learn that as a kind of bonus, the future looks bright. "The twenty-first century is going to see astounding changes in everything, including the role of violence" (10).

It is true that, understandably given his background, Morris is more focused on modern humans than on our distant ancestors, more on history than on evolution. Some might not think this enough, for biology-culture clash is far from over, with a group of practitioners strongly and publicly reaffirming the standard social science line. This is the so-called "Seville Statement on Violence," 1986.

> It is scientifically incorrect to say that we have inherited a tendency to make war from our animal ancestors. Although fighting occurs widely throughout animal species, only a few cases of destructive intra-species fighting between organized groups have ever been reported among naturally living species, and none of these involve the use of tools designed to be weapons. Normal predatory feeding upon other species cannot be equated with intra-species violence. Warfare is a peculiarly human phenomenon and does not occur in other animals.

They continue that the "fact that warfare has changed so radically over time indicates that it is a product of culture." Anything biological like language has a purely support role. When one learns that UNESCO adopted this declaration at the twenty-fifth session of the general conference on November 16, 1989, one can only assume that the body of its first director-general Julian Huxley is revolving rapidly in its grave. With relief, we don't have to resolve this conflict. It is enough here to point it out—although we shall be returning to its message—and to mention that differences like this are precisely what one expects in religions.

We have arrived at the new millennium. We have two rival perspectives on the human state—Christianity and Darwinism—and not only are these perspectives religious or religious-like, but they show this in the ways they treat the problem of war. More than this, it is a toss-up as to who is the more important in this story: St. Augustine or Charles Darwin. There is a huge amount of overlap in thinking about our original, innate nature, and there is good evidence suggesting that it was the ideas of the saint that filtered through into the ideas of the biologists. This can be said, notwithstanding the very important point that for both Christians and Darwinians there have always been dissenters from the norm, people who simply don't accept that we are innately violent. Some do this because of their Christianity. Some do this, whether they be believers like Allee and Thorpe, or nonbelievers like Montagu and the Seville signers, because of their interpretations of the physical evidence. As in the majority position, one suspects that the religious non-Augustinian take on human nature has been an influence over in the science and its supposed implications, secular or sacred.

For now, let us leave both the majority and the minority positions, and turn first to completing our story. This we do in the next chapter, taking up two recent pertinent works on our topic, representing our two poles.

11 | Rival Paradigms

THIS CHAPTER IS FRAMED BY two recent books on war, one by a Christian and the other by a Darwinian.

In Defence of War

Nigel Biggar (as noted in the *Prolegomenon*) is the Regius Professor of Moral and Pastoral Theology at the University of Oxford, where he is also a Canon of Christ Church Cathedral. In both of these roles, Biggar follows in the footsteps of another Anglican priest, Oliver O'Donovan. A predecessor also interested in war who argued, at the beginning of this century, in *The Just War Revisited* (2003), that the growth of culture demands war, that turning against war is "no demand of natural law" (5), that "armed conflict can and must be re-conceived as an extraordinary extension of normal acts of judgment" (6), and for good measure, in the tradition of Augustine and Thomas, that in "the context of war we find in its sharpest and most paradoxical form that love can sometimes smite, and even slay" (9). In the mold, in his work *In Defence of War*, writing as a Christian—as an Anglican—Biggar offers another unequivocal defense of a conservative, Augustinian take on just war theory. He believes that war is a dreadful evil. He believes that humans are tainted by original sin. He believes that humans may legitimately take up arms in the right cause. For good measure, he believes that Darwinism offers a false philosophy that should be rejected. Thus, he writes: "there is every reason to avoid war, of course, if war may be avoided. The evils it causes are indeed terrible." Surely, there "has to be a better way to respond to grave injustice than by waging war"? Could we not engage with our opponents, "reasoning with them sympathetically and patiently, even to the extent of addressing their legitimate grievances"? Alas! "It seems empirically self-evident to me that not all human beings are well motivated or well intentioned—and it seems so as much from internal reflection as from external observation." The darker Augustinian psychology rings more truly than the naïve belief in human nature that we associate with Rousseau. For all our rationality, "we are also—and primarily—creatures of love or desire; and when desire throws off reason's control, we become creatures of passion; and when we are creatures of passion, we lose the ability to contain ourselves out of respect for justice for others." In short, war is nigh inevitable in the sense of empirically bound to happen. "I do not believe that there always has to be an available pacific solution" (Biggar 2013, 9).

Biggar has little time for pacifism, denying it on traditional grounds both biblical and by appeal to natural law. He shares the eschatological commitment of the pacifist: "To believe in the God who wears the face of Jesus is to believe in a superhuman power that is at work in the world to recover it from sin and deliver it to fulfilment. It is to believe that the world's salvation does not lie entirely, or primarily in human hands; and that it does not lie wholly in the future, but is already coming to be" (20). Note the Augustinian amillennialism here. We are in a sense already in end times. Amillennial perhaps rather than premillennial, but there is no place here for progress. Also with Augustine there is a non-pacifist reading of the Sermon on the Mount. "The New Testament does forbid certain kinds of violence: that which is disproportionate because motivated by contemptuous or hateful or vengeful anger" (49). However, "its prohibition of violence is specific, not absolute. Therefore it is not accurate to summarize the New Testament's position in terms of a commitment to 'non-violence' *simpliciter*." Augustinian in a way that Reinhold Niebuhr is not. The God of the New Testament allows and expects us to fight. It is not that we are caught in the bind that as private individuals we should not fight, but as public individuals we must fight.

Behind all of this lies original sin—"the fated dimension of human choices under the weight of history's socio-psychological legacy. If we are free, we are free only within bounds; and the bounds are unequal, for history has dealt more kindly with some than with others." Again, deeply Augustinian, including that the saint allowed God had been more bountiful to some than to others. Take note of the extent to which this kind of thinking goes against a non-Augustinian like Kant, who insisted on the absolute freedom of every human being. Little wonder now that Biggar thinks we can make the case for the religious legitimacy of just war theory: "it may be permissible to choose to act in such a way as to cause the death of a human being, provided that what is intended is something other than his death (e.g. defending the innocent), that the possibility (or even certainty) of his death is accepted with an appropriate and manifest reluctance, and that this acceptance is necessary, non-subversive, and proportionate" (101). "Proportionate" is the weasel word here. Take the battle of the Somme, the first day of which—July 1, 1916—saw nearly sixty thousand British casualties, of whom twenty thousand were dead. It took the pressure off the French and drained the Germans, who ended the battle with at least half a million casualties. This leads Biggar to reflect: "The Somme was costly—excessively costly—but it was not futile, and nor were its achievements trivial. It is not clear, then, that the Somme was not worth 622,000 Allied casualties. Indeed—though I tremble to say it—it is not clear that it would not have been worth many more" (129).

You might think that someone who reasons like this is closer to Friedrich von Bernhardi than anyone else we have encountered in this book. Not so, for Biggar argues strongly that Germany started the Great War and it, and it alone, is responsible. Someone like von Bernhardi (who is not mentioned by name) saw war as a means of aggression, of making sure of your place in the world and if need be taking the places of others. Just war theory speaks rather to defense of the right in the face of the wrong. Biggar takes this as an opportunity to put the boot into the Darwinian position! "But if social Darwinism thinks it natural for a nation to launch a war simply to prevent the loss of military or diplomatic dominance, just war doctrine does not think it right. Just cause must consist of an injury; and Germany had suffered none" (135). Adding later: "It is true that Germany launched war against France because it felt itself surrounded by enemies and believed that its very survival depended on fighting sooner rather than later. It saw its aggression as a form of anticipatory defence. However, whereas social Darwinism might sanction such a view, just war theory does not" (211). There was no reason to think that France and its allies, Russia and Britain, were about to attack Germany. There was no legitimate reason for preemptive behavior at this point. Indeed, "French forces were ordered to keep ten kilometers behind the frontier, so as to avoid communicating offensive intentions" (211).

Expectedly, in the light of German-like behavior, there isn't much love of Kantian notions about perpetual peace and that sort of naïve optimism. Biggar would give Hegel a run for his money. Talking of those who put their hopes in organizations like the United Nations: "I respect the aspiration; but I cannot share the faith. My Protestant view of human community and institutions will not allow it. I do not think we can expect human society—even as it manifests itself in international courts—to be free of moral disagreement and political conflict" (241). Somewhat sniffily he adds that this is the kind of erroneous thinking we find with Catholics. The authority of their Church having collapsed during the Reformation, they hope to regain it or substitute for it; adding that "my Inner Protestant regards the church and the UN with equal skepticism" (241, fn 81). Biggar would think that he has as much right to turn to Augustine for guidance and inspiration. The Bishop of Hippo is venerated as much if not more by Protestants as by Catholics. It is true that a hundred years before Anglican theologians had rather lost sight of this, but if you go back to the Thirty-Nine Articles themselves, the Augustinian content is loud and clear. "It is lawful for Christian men, at the commandment of the Magistrate, to wear weapons, and serve in the wars."

Bringing the story up to date, offering an Anglican contribution to the Iraq problem, complementing William V. O'Brien's earlier (1992) defense of Desert Storm, Biggar then goes on to defend the Bush-Blair invasion of Iraq in 2003. "For sure, the invasion against Saddam Hussein suffered from some serious flaws." These included "excessive optimism, impatience with human frailty, managerial over-confidence, heedlessness of uncongenial counsel . . ." Nevertheless, "I judge that the invasion of Iraq was justified" (325). Again, the Augustinian themes are up front. We are assured that: "As a considerably Augustinian Christian, rather than a Hobbesian atheist, I am not cynical about human beings." More realistic. "I do think that Augustine's biblically inspired insight into human psychology is far more deeply illuminating that than attributed to the ancient Greeks. The cause of our wrongdoing is not always or basically ignorance, but ill-ordered *love*" (328). Most of us most of the time would be very happy to live well-regulated peaceful lives. It is just that human nature keeps welling up and making it impossible. As a parting shot, Biggar complements this dark vision of humankind with another blow at the rival, optimistic, Darwinian vision of humankind. Just before Biggar's book was published, the Darwinian linguist Steven Pinker—as also noted, Johnstone Family Professor in the Department of Psychology at Harvard University—had made the case for the other side. Unfortunately, this book *The Better Angels of Our Nature* is "little more than well-meaning speculation." Biggar agrees with a reviewer who wrote: "Pinker's book is a personal expression of moral idealism but not, alas, an account of historical reality" (Biggar 2013, 329). If more amillennial than strictly premillennial, Biggar knew a postmillennialist when he saw one, and he didn't much like them.

The Better Angels of Our Nature

Let us complement our discussion by turning at once to Pinker, someone as distinguished in the Darwinian world as Biggar is in the Christian world. *The Better Angels of Our Nature* is a long book, but the central thesis is simple. We humans are a violent folk—"most of us—including you, dear reader—are wired for violence, even if in all likelihood we will never have an occasion to use it" (Pinker 2011, 483). Continuing that although it is true that "when men confront each other in face-to-face conflict, they often exercise restraint. But this reticence is not a sign that humans are gentle and compassionate. On the contrary, it's just what one would expect from the analyses of violence by Hobbes and Darwin" (487). We want to kill the other fellow but we are also aware that he wants to kill us, so we take care. Don't rush into it. Take your time. Restrain yourself until the signs look favorable. Nothing Christian here, just evolutionarily produced common sense.

Yet for all the inclination to violence, over the years we have restrained ourselves and today in fact we are a lot less violent than we used to be. "Believe it or not—and I know that most people today do not—violence has declined over long stretches of time, and today we may be living in the most peaceable era in our species' existence" (xxi). This has had significant consequences. "Daily existence is very different if you always have to worry about being abducted, raped, or killed, and it's hard to develop sophisticated arts, learning, or commerce if the institutions that support them are looted and burned as quickly as they are built." How has this all come about? The fact of the matter is that, for all of our violent nature, we are a mixed bag. Sometimes violent; sometimes not. This is the legacy of our evolution. The question is how the nonviolent side has gained the upper hand. "The way to explain the decline in violence is to identify the changes in our cultural and material milieu that have given our peaceable motives the upper hand" (xxiii).

Pinker identifies six key trends. Note that these are all very recent in the overall evolutionary time scale of things, so it is more a matter of a relatively sudden, culture-fueled shift than of even a relatively rapid evolutionary change like the arrival of our large brains. The first trend, which Pinker calls the "Pacification Process," involved the move from hunting and gathering to agriculture. No longer were raiding and fighting the keys to success. It was rather a matter of getting on with the job at hand and focusing on what you have and making it better. Second came the "Civilizing

Process," something happening over the past five hundred years or so. It involved a rapid drop in violence, homicide particularly, and is linked with the rise of the modern state, with a sophisticated infrastructure and commerce and so forth. Third, there is the "Humanitarian Revolution." "It saw the first organized movements to abolish socially sanctioned forms of violence like despotism, slavery, dueling, judicial torture, superstitious killing, sadistic punishment, and cruelty to animals, together with the first stirrings of systematic pacifism" (xxiv). Next, fourth, comes the "Long Peace." This refers to the way in the last century how, after two bloody world wars, the major powers stopped fighting each other. Fifth, even more recently we have the "New Peace." "Though it may be hard for news readers to believe, since the end of the Cold War in 1989, organized conflicts of all kinds—civil wars, genocides, repression by autocratic governments, and terrorist attacks—have declined throughout the world." Finally, sixth, we have the "Rights Revolution." Pinker refers here to a "growing revulsion against aggression on smaller scales, including violence against ethnic minorities, women, children, homosexuals, and animals" (xxiv).

As he has said, many of these claims are counter intuitive, so most of the book is devoted to examining the evidence. Thus, for instance, in starting the discussion on pacification, Pinker quotes the philosopher Thomas Hobbes.

> So that in the nature of man, we find three principal causes of quarrel. First, competition; secondly, diffidence; thirdly, glory. The first maketh men invade for gain; the second, for safety; and the third, for reputation. The first use violence, to make themselves masters of other men's persons, wives, children, and cattle; the second, to defend them; the third, for trifles, as a word, a smile, a different opinion, and any other sign of undervalue, either direct in their persons or by reflection in their kindred, their friends, their nation, their profession, or their name. (Pinker 2011, 33, quoting *Leviathan*, 1651)

Chimpanzees back up this innate propensity to violence. When Jane Goodall started her studies, no one believed her when she recorded their violence. Or, they thought she must be dealing with abnormal circumstances. Perhaps it is all a function of primatologists leaving them food to eat. "Three decades later little doubt remains that lethal aggression is part of chimpanzees' normal behavioral repertoire" (38). At least fifty animals have been killed in inter-group conflict, and another twenty-five in intra-group conflict. These deaths were spread out over several (at least nine) communities, and they included entirely natural groups—never provisioned or otherwise aided or disturbed by humans. "In some communities, more than a third of the males die from violence" (38).

Back to humans, and to sexual selection with males competing for mates. "The more recent and abundant *Homo* fossils show that the males have been larger than the females for at least two million years, by at least as great a ratio as in modern humans. This reinforces the suspicion that violent competition among men has a long history in our evolutionary lineage" (40). And yet, once you start to put the figures together, you see that there has been a rapid and major decline in warfare and subsequent deaths. The move to forming states has had huge consequences. It simply isn't reasonable to think that this is all a matter of pure chance. Nor need you, for if you analyze the evidence in more detail, you find that suppositions hold true. In one major study: "The proportion of hunter-gatherers that showed signs of violent trauma was 13.4 percent" (51). Conversely: "The proportion of city dwellers that showed signs of violent trauma was 2.7 percent. So holding many factors constant, we find that living in a civilization reduces one's chances of being a victim of violence fivefold." The root cause is so obviously government control that most anthropologists simply don't bother to record the fact. "The various 'paxes' that one reads about in history books—the Pax Romana, Islamica, Mongolica, Hispanica, Otomania, Sinica, Britannica, Australiana (in New Guinea), Canadiana (in the Pacific Northwest), and Praetoriana (in South Africa)—refer to the reduction in raiding, feuding, and warfare in the territories brought under the control of an effective government" (55–56).

We keep going through the categories. With respect to the Civilizing trend, there has been a 95 percent decline in homicides in England, despite people's general assumption that we today are a more violent society than was the case in the Middle Ages. The Humanitarian Revolution speaks of such things as the decline of slavery. "Some of this progress—and if it isn't progress, I don't know what is—was propelled by ideas: by explicit arguments that institutionalized violence ought to be minimized or abolished." Some of the progress "was propelled by a change in sensibilities. People began to sympathize with more of their fellow human beings, and were no longer indifferent to their suffering. A new ideology coalesced from these forces, one that placed life and happiness at the center of values, and that used reason and evidence to motivate the design of institutions" (133). Also stressed is the importance of writing and literacy. "The pokey little world of village and clan, accessible through the five senses and informed by a single content provider, the church, gave way to a phantasmagoria of people, places, cultures, and ideas" (174). This in itself makes for people behaving in nicer ways toward each other. Progress! Note the jab at the church. We have progress, but it seems that at least part of that progress involves breaking the hold of organized religion. As is so often the case, the Darwinian picture is pitched in direct (rather hostile) opposition to the Christian picture.

Pinker moves us toward peace and the end of war. Kant is given a star billing. Kant's moral philosophy highlighted the supreme ethical principle—the Categorical Imperative—given in various formulations as the need to act only in ways that could be universalized or as the need to treat others as ends in themselves and not simply as means to other ends. "Could the Long Peace represent the ascendency in the international arena of the Categorical Imperative?" (292). Elaborating: "Norms among the influential constituencies in developed countries may have evolved to incorporate the conviction that war is inherently immoral because of its costs to human well-being, and that it can be justified only on the rare occasions when it is certain to prevent even greater costs to human well-being" (291).

Having laid out his case, Pinker turns then to causal analysis. He looks first at our "inner demons," the forces that make for human violence. We have seen something of this already. It is what you expect in a Darwinian world. To reproduce, you must succeed, and to succeed you must be willing to fight. The struggle for existence may often be a metaphor, but not always. Pinker explains that violence is not a kind of hydraulic urge that bursts out automatically when it gets full or under too much pressure. It is a matter of many different things. He lists instrumental violence—violence simply to get some thing or end; dominance—getting to the top spot and controlling others; revenge—obviously, getting your own back (and with obvious Darwinian roots since people will know that this is likely to happen); sadism—enjoying the suffering of others; and (a little different) ideology—something involving violence to the end of a utopia, an unlimited good. It is, to be candid, not easy to see how all of these fit the adaptive pattern, but Pinker is always willing to give things the college try. Take sadism. There are various possible motives including "a morbid fascination with the vulnerability of living things, a phenomenon perhaps best captured by the word macabre. This is what leads boys to pull the legs off grasshoppers and to fry ants with a magnifying glass. It leads adults to rubberneck at the scene of automobile accidents—a vice that can tie up traffic for miles—and to fork over their disposable income to read and watch gory entertainment" (549–550). All of this apparently could be motivated by "mastery over the living world, including our own safety." One explanation, I suppose, for the gratification so many derive from watching the violence of American football.

Complementing our inner demons are our "better angels." Pinker identifies four. Empathy—feeling the pain of others and making their interests at one with ours; self-control—anticipating the consequences of actions and controlling these actions on the basis of the anticipation; moral sense—having norms and taboos that govern and reduce violence between people; and reason—being able to pull back and look at and think about our and the general situation. One big question is whether the ways in which our various drives come together to form the trends discussed earlier are all ultimately a matter of culture, or could it be that genetic changes are also involved? Are the better angels conquering the inner demons purely because of environmental and other triggers, or are we actually in a Darwinian biological sense moving toward being nicer people? This is not a silly question because we are becoming very much aware of how quickly and decisively natural selection can act, including act on human nature. Pinker's conclusion is that "while biological evolution may, in theory, have tweaked our inclinations towards violence and nonviolence, we have no good evidence that it actually has. At the same time, we do have good evidence for changes that could not possibly be genetic, because they unfolded on time scales that are too rapid to be explained by natural selection, even with the new understanding of how recently it has acted" (621). By rights, since so many criminals were exported to Australia, today it ought to be one of the most violent nations on the planet. However, for all that Aussies present themselves as tanned he-men, interested only in sport, suds, and sheilas, the facts tell otherwise. Their murder rate is lower than that of the home country, and on an absolute scale is one of the lowest in the world. Bigger softies than the Mongolians, apparently.

We come to the end. How has the decline in violence come about? Five factors are highlighted: Leviathan—having a state that has taken to itself the only legitimate use of power and force in itself reduces violence; Commerce—trading with others means you have a purely selfish interest in their staying alive and in their general well-being; Feminization—clearly this takes you away from the cult of violence so beloved of young men and instills other values in people; Cosmopolitanism—things like the rise of literacy broadens horizons and makes people more understanding of and sympathetic to the nature and needs of others; and finally Reason—this leads us to see the pointlessness of violence and how we all benefit if it is reduced.

Pinker opens his final chapter with a quote from an earlier writer.

> As man advances in civilisation, and small tribes are united into larger communities, the simplest reason would tell each individual that he ought to extend his social instincts and sympathies to all the members of the same nation, though personally unknown to him. This point being once reached, there is only an artificial barrier to prevent his sympathies extending to the men of all nations and races. (671)

We have come full circle. We are back with Charles Darwin in the *Descent of Man*!

Rival Visions

The problem I set out to solve at the beginning of this book was the controversial status of Darwinian thinking today. Why is it that so many people find Darwin's legacy so unacceptable, not merely false but in some sense radically misconceived and dangerous? The simple answer is that it goes against religion; in particular, it goes against the Christian religion. The findings of this book support this answer. Yet, this cannot be the whole story. There are other areas of science that sit uncomfortably with religion and yet the two sides manage to get along reasonably well. To me, it beggars belief to think that Planet Earth, this speck of dust in a vast universe, can really be so unbelievably important that it uniquely, and its human denizens in particular, can be the scene of the divine drama concerning the Christ and our redemption. We, and we alone, have won the lottery—or, at least, the right to enter the lottery. And that is before you get to multiverses. Yet, you don't get today's Christians tearing their hair out at modern astronomy. Andromeda is not an obsession of those Protestant pastors of mega-churches across America. Conservative politicians do not spend their days trying to impose the Old Testament view of the universe on the physics classes in public schools in their states. There are no museums to geocentrism. What is it about Darwin's legacy that so upsets those pastors? What is it about the legacy that led Steven Jay Gould and me and Francisco J. Ayala to Little Rock, Arkansas, in 1981, where we appeared as expert witnesses for the ACLU in that earlier-mentioned (successful) lawsuit against a new state law mandating the "balanced treatment" in teaching about origins—Darwin versus Genesis? What is it about the legacy that spurred the building of a Creationist Museum—complete with monstrous facsimile of Noah's Ark—just outside Cincinnati? My answer and solution lies in Darwinian thinking being more than straight professional science—although it is certainly that and very successfully so. I argue that there is a side to Darwinian thinking, what I refer to as Darwinism, that functions as a religion, or if you prefer, a secular religious perspective. Those pastors, those politicians, those museum builders know a rival when they see one.

I chose the topic of war as a case study, and I claim now that I have confirmed my hypothesis. At least, I have offered solid evidence for my hypothesis. It is only a case study, and so the full explanation of the opposition to Darwin's legacy must be much more comprehensive, going across a host of issues like moral behavior, the teaching of our children, the punishment of evil doers, the place and status of women and minorities, and far more. The intellectual godfather of Intelligent Design Theory, retired law professor Phillip Johnson, is obsessed at the way in which he thinks Darwinism promotes feminism—"stroppy broads in pantsuits," a phrase I have heard used—away from the proper biblically sanctioned role of women in society (Johnson 1995). Frankly, on reading the *Descent*, I am not sure about this, but I take the point. Certainly, I see Darwinism as having a broad range across the issues that concern Christians. Richard Dawkins in the *God Delusion* (2006) holding forth about the evil intent behind the bringing up of children in the Christian faith. Jerry Coyne, retired from the biology department at Chicago and leading expert on speciation, in *Faith vs Fact* (2015) giving us his views on free will—there isn't any!—and (especially in his blog, "Why evolution is true") its implications for crime and punishment. Edward O. Wilson in *Biophilia* (1983) and many other books with his obsessive concerns about how we are destroying the environment and having horrendous effects on the planet's biodiversity. About our Christian complacency in thinking that God will step in and put things aright.

Darwinism on war is a case study, but it is an important case study. The basic (majority) picture at the heart of Christianity is that we are beings made in the image of a good God, but that we are fallen and hence are tainted by sin. Because God loves us, he came to us in the form of Jesus and died on the cross. For whatever reason, we now have the possibility of salvation and eternal bliss with our creator. Nothing we can do merits this. We alone cannot bring on meaningful or lasting improvement. We are in the hands of God. Providence. Darwinism tells a very different story, but like traditional religion—no great surprise since it is in many respects an offspring of Christianity—it speaks to human nature and to our obligations and to the future. We are the end products of the cruel, blind process of natural selection. What happens was due to the laws of nature, and we have no reason to think that we have any spiritual status or that there is a god who in any sense cares for us. However, all is not without meaning. We started as brutes, but now humans have reached the pinnacle of evolution, and there is hope of further advance be it biological or cultural. The title of Steven Pinker's newest book could not be more revealing or confirming: *Enlightenment Now: The Case for Reason, Science, Humanism, and Progress* (2018). Progress! In eschatological terms, the options are putting it all in the hands of God versus rolling up our sleeves and getting on with the job ourselves.

The rival treatments of war—the majority rival treatments of war—fit right into these two world pictures. For the Pauline Christian, because we are sinners, war is going to happen. This is a fact and one we cannot avoid. The question is what should we do in the face of this? Whatever it is, it must be such as to fit into God's will. There are times when we must fight evil, when it is our Christian duty to go to war. There is going to be no war to end all wars. We must nevertheless do our best, and fight within constraints imposed by our God. For the Darwinian, there are good reasons why war was a key causal part of our evolution. It is the most obvious manifestation of the struggle for existence. In some sense, therefore, because war led progressively up to humankind, war has to be thought a good thing. However, that time is now past. We have evolved so we ourselves have the ability to limit or even to end war. This we must do. The end of war is an achievable aim, even though it be very difficult. These claims may be based on science, but they are not in themselves scientific claims. They are religious claims or claims made within a religion. They are claims that start with a vision of human nature that almost certainly owes as much to St. Augustine as does the Christian position, given then a scientific gloss. Darwinism.

In the *Prolegomenon*, I defended the moral worth of this inquiry and I stand by this. I acknowledged, however, that deliberately I was arguing within self-imposed constraints, particularly about the truth values of much I was discussing. It is not always easy to do this, and there are times when one feels that, whatever the argument, one should not do this. I, for one, find it incomprehensible that responsible churchmen like the Bishop of London, Arthur Winnington Ingram, were not crying bloody murder by the time of the Battle of the Somme in 1916. I simply do not know how William Temple, hugely revered, future Archbishop of Canterbury, slept again given the death of 355 of his boys on the battlefields of that totally pointless war. The views of his successor, Archbishop Geoffrey Fisher, are best passed over in silence. Christians are not alone. If only by courtesy allowing him the title "Darwinian," equally incomprehensible (and reprehensible) is the dreadful Friedrich von Bernhardi who was pleading the moral virtues of war as ardently in 1918 as he was so doing in 1914. Vernon Kellogg had a point about the kind of stuff he was spouting. At the same time, going the other way, much credit must go to Kellogg himself, who showed true greatness in his work in the Great War and in his ever-sensitive writings on the subject. Likewise, one cherishes the memory of Father John C. Ford who unambiguously showed the grave immorality of the Allied obliteration bombing in the second half of the Second World War. Harry Fosdick is a noble addition to the American tradition of great Protestant pastors and preachers. In a similar vein, there is yet more reason for regarding with huge respect and affection that avowed Darwinian William James, seeking desperately for moral and worthwhile outlets for the more dangerous aspects of human nature, especially the human nature of the young and idealistic. The awful thinking sticks more in the mind than the good thinking, but they are both there.

Yet, speaking overall, worrying too much about truth stands in the way of what I wanted to do. I am very much not in the business of proving one side better than the other. Very quickly, the original quest would be lost and forgotten, and I would be playing the same game as those whom I criticize. I acknowledge also though, that the constraints do limit the nature and value of my conclusion. If we are indeed faced with rival religions, the question then becomes—what next? Even if we are not—as I certainly am not—interested in converting everyone either to a form of Augustinian Christianity or to Darwinian agnosticism or Dawkinsian atheism, is there some way of bringing the two positions closer together? Even if they are not to be blended, could they be brought more in harmony or in parallel so the conflict is reduced? Perhaps even could one hope for an appreciation of the other side, and some level of cross-fertilization? Even though extremists at both ends will not be interested, might there be fertile and profitable work for those of us more in the middle, more accommodating to use a term that some hurl with scorn but that others adopt with pride?

In one sense, you may think it does not seem particularly promising. Grant that Darwinism, as understood in this book, takes on the role of a secular religion. It does still seem that the interests of the Christian and the interests of the Darwinian are in different directions. For obvious reasons, Christians are particularly interested in the moral and political aspects of war. When can one legitimately wage war and under what circumstances? For obvious reasons, Darwinians are particularly interested in issues that are more factual. What leads people to war and how has it varied through history? But already one starts to see places of overlap and possible mutual interest and understanding. Just war theory stresses the importance of respecting noncombatants—the innocent, the uninvolved, the young, the old, the sick, and so forth. Darwinians are much interested in which members of an opposing group are going to be the focus of ire and hostility. Will a marauding band always try to wipe out all of the opponents and if so why, and if not, why not? What about protecting the young of one's opponents, especially if they are female? There is surely room for at least some shared inquiry here.

More generally, Christians and Darwinians share an obsession with human nature, and this indeed leads to the paradox that underlies our whole discussion. Christians think us tainted with original sin and, for this reason, although war is always bad, we will have it with us always. Morbid fatalism. Darwinians think we are primates who evolved through natural selection. Violence is innate but it was highly adaptive. For this reason, war in the past has to be judged sometimes good, and, at the same time, thanks to the powerful new adaptation of culture, there is the hope of an escape from our heritage. Naïve optimism. Could it not be that thinking again about human nature—or rather rethinking human nature in the light of what we now know—might help to break this paradox by showing that neither side is or can be right? They are both mistaken about human nature and, when we see this fact, can we see also the chance to move forward together more fruitfully? Could it be that our minority positions, Christian and Darwinian, have something to tell us? I am not now suggesting that everyone must become a pacifist, but that their alternative thinking (or thinkings) about human nature might have something to tell us. Let us conclude this book by turning to these questions.

12 | Moving Forward

Christian Human Nature

START WITH CHRISTIANITY AND WITH the vexed issue of original sin. The standard Augustinian picture within which we have been working has humans created good—"in the image of God"—but then in some sense falling away. After some speculation (to be found in the non-canonical Book of Enoch) that the problems might have started with the illicit sexual union of apostate angels and humans (as supposedly detailed in Genesis 6), general opinion settled on the sin of disobedience by Adam and Eve in the Garden, their subsequent expulsion, and the taint for all of their descendants of that original bad act (Williams 1927). All humans are without exception (possibly the Virgin) born with an inclination to do wrong. "Human nature was certainly originally created blameless and without any fault; but the human nature by which each one of us is now born of Adam requires a physician, because it is not healthy" (McGrath 1995, 219, quoting Augustine, *On Nature and Grace*). God therefore sent Jesus—wholly God and wholly human—for our salvation. "This came through his death on the cross. He was a sacrifice. For whereas by His death the one and most real sacrifice was offered up for us, whatever fault there was, whence principalities and powers held us fast as of right to pay its penalty, He cleansed, abolished, extinguished; and by His own resurrection He also called us whom He predestinated to a new life; and whom He called, them He justified; and whom He justified, them He glorified" (Augustine 1886, *On the Holy Trinity*, 4.13.17). All that was needed now was for someone to explain why the sacrifice was needed and how it was that only Christ could do the job. St. Anselm obliged. "It is necessary that God should fulfill his purpose respecting human nature.

And this cannot be unless there be a complete satisfaction made for sin; and this no sinner can make. Satisfaction cannot be made unless there is someone who is able to pay to God for the sin of humanity. This payment must be something greater than all that is not God, except God" (McGrath 1995, 182–183, quoting Anselm, *Cur Deos Homo*).

Note that right at the heart of this picture is the claim that we are in a mess because of what we—Adam and Eve—have done, and hence we are in need of forgiveness and excuse and redemption or whatever. We are sinners and action is needed to put this right. Now the problem! Modern science, modern Darwinian paleoanthropology, says that this story is fundamentally and irretrievably mistaken (Ruse 2012). It is false. There was no original couple. There was no Adam and Eve, the unique parents of all humans and the reason why we are sinners. Humans leaving Africa may have gone through bottlenecks—the lack of variability suggests that we almost certainly did—but overall there were never less than about ten thousand humans. Ten thousand Adams and Eves, and before you suggest that one of the Adams might have been the grandfather of us all, this is probably so, but so also is the fact that there were almost certainly hundreds of other like Adams and Eves—all of whom were our grandparents. More than this, it is simply not possible that Adam and Eve were perfect, never sinning until that fateful afternoon when they ate the apple. They were humans from the start—nice and nasty—and so were their mums and dads, and uncles and aunts, and grandparents all the way back to the monkeys. They weren't too perfect either!

The Augustinian human simply doesn't exist. At once, we have removed the foundations of the traditional Christian picture of war. War may be highly probable—or not. It is not inevitable, or even contingently inescapable, because we are not tainted in that way. Kant, who thought on good philosophical grounds that we should be responsible for our own actions and not put it all on Adam, objected to the Augustinian notion of sin. He spotted the pertinent possibilities. "For man, therefore, who despite a corrupted heart yet possesses a good will, there remains hope of a return to the good from which he has strayed" (Kant 1793). Obviously, one would have to be obtuse to deny that in the past century humans have done some dreadful, dreadful things that have led to war and suffering. Any species that produces Heinrich Himmler knows altogether too much about sin. That is another matter. It is the predictive certainty that is at issue here. Of great issue, for if sin is no longer predictively certain, then there is the prospect of improvement and peace. You may be cynical and think this unlikely; but, if nothing else, you are no longer caught with almost complacently expecting and putting up with war. To requote the philosopher Robert Holmes: "While one cannot know what would have taken place had Christianity not charted for itself a new course after Augustine, the consequences of the course it did take holds increasingly grim prospects as they unfold in our modern world. It is as though belief in the pessimistic picture painted by Augustine has helped bring about the truth of that belief" (Holmes 1989, 145). When one thinks of Biggar on the Somme and Iraq, he may have a point.

Am I now saying that the only way we get out of the traditional Christian take on war is by giving up Christianity? Absolutely not! One counter move is to go on defending the Augustinian position by insisting that no sophisticated Christian thinker today takes literally the story of Adam and Eve. An obvious way of executing this defense is by appealing to evolutionary biology. Inherited sin is not something that came about through the act of Adam but something that comes with our biological nature. We are sinners because, as Thomas Henry Huxley pointed out, the things that lead to sin are good adaptations for survival and reproduction. This not only explains sin, but shows why it is very unlikely that we will get rid of it once and for all. Even if culture can overcome sin, cultures change, and reverses occur. From the Weimar Republic to the Third Reich.

Of our sinful behaviors, evolutionary biologist Daryl P. Domning (2001) writes: "it is demonstrable by experiment and fully in accord with Darwinian theory that these behaviors exist because they promote the survival and reproduction of those individuals that perform them. Having once originated (ultimately through mutation), they persist because they are favored by natural selection for survival in the organisms' natural environments." This explains the "stain of original sin." Culture is involved but ultimately it is biology.

> Indeed, we do learn to sin from the sinful society into which we are born; and even the very first humans learned to sin from the selfish though sinless pre-human society into which they were born. But even without that legacy of learned behavior, we would still be urged to sin by the genetically programmed selfishness, dating from the dawn of life, that underlies it and gave rise to it.[1]

[1] Candor compels me to admit that this line of argument has been promoted by an author close to my heart. "Original sin is part of the biological package. It comes with being human. We inherit it from our parents and they from their parents: they acted as they did, and because they acted as they did, it is passed down to us" (Ruse 2001, 210). None of us are perfect, a reflection that—for all of

Leaving for a moment the empirical claims being supposed here, there is a major theological flaw that undermines this defense. If our sinful nature is no longer the result of a free act, but part of our inherited nature, then its ontology has been shifted from something external to but imposed on human nature, to being fully part of human nature. No need of the Catholic "original righteousness." No need of Protestant worries about which parts of human nature are those forced on us by sin. Now, none of this in itself stops us from being made in the image of God—after all, humans have penises and vaginas and these are very much not part of the Judaeo-Christian God—but it does make ineffective the sacrifice on the Cross. The sacrifice is for something imposed on God-created human nature, not for something always part of God-created human nature. The death on the Cross cannot suppress our appetites. No more can it suppress our violent, biological nature. The biology move is theologically flawed. Which it has to be, for we are now saying that our violent, biological nature is part of our God-created human nature.

Another way of executing the defense is that taken by Reinhold Niebuhr. He argues that the Fall is not literal but symbolic. "The myth does not record any actions of Adam which were sinless, although much is made in theology of the perfection he had before the Fall" (Niebuhr 1941, 279–280). Continuing: "Adam was sinless before he acted and sinful in his first recorded action. His sinlessness, in other words, preceded his first significant action and his sinfulness came to light in that action. This is a symbol for the whole of human history." Robert Merrihew Adams (1999) suggests that this puts Niebuhr (perhaps not consciously) in a tradition started by Kant. I doubt this. Unlike Adams, I am loathe to allow that the non-Augustinian Kant could still have a theory of original sin. I would agree however that Niebuhr shares with Kant the belief that sin (however it is called) is a matter of personal failings and not to be referred back to a dubiously existing ancestor. Again for the moment putting on one side the empirical, I argue that it is precisely this agreement that leads Niebuhr into a theological swamp. Two problems spring at once to mind. First, if Adam is symbolic for us all, then it seems that all of us have a pre-sin time when we make that first bad choice. Original sin can apply only to our actions after this. Which rather suggests that we are being plunged into the Pelagian heresy, because apparently we all have total freedom to choose good or ill. The window of opportunity is limited, but it is there. Second,

---

my nonbelief—puts me more firmly in the Christian than the Darwinian tradition. I trust by now the reader realizes, although I have major disagreements, I do not take this as a slur.

it is all very well talking about symbolic, but symbolic of what? A cigar is symbolic, but only in the sense that it represents a penis. Remember, sometimes a cigar is just a smoke. No penis, no symbol. Given the context and the time, one suspects that Niebuhr has some psychological explanation in mind. We sin and, in some sense, this turns us to sin. The slippery slope.

Again though, we start to wonder about how it all came about. If we are going to take Freud and his followers seriously, we know that ultimately it is all going to end with biology. A point noted by Niebuhr's student, the Chicago theologian Langdon Gilkey. "Niebuhr seemed quite unaware of the great influence of modern science on his theological understanding." Continuing that it is "largely because of the new understanding of geological and biological science of nature's evolutionary past and of archeology and anthropology of the human past that Biblical narratives about origins were transformed from narrations about particular historical events, about our 'first ancestors,' into narrations containing symbolically important theological content: the good creation, the temptation and the fall, the expulsion and subsequent suffering" (Gilkey 2001, 234). Which takes us right back to above-expressed worries about all of these things being predicated on our essential nature and hence not subjects for the Atonement.

Apart from modifying or tempering the Augustinian position, which could involve not just working on the meaning of Genesis but wriggling a bit about what modern science tells us—Notre Dame theologian Celia Deane-Drummond (2017) makes much of the evolution of an awareness that we are doing wrong, as opposed to merely the evolution of unfriendly traits—the second counter move is to recognize what our minority Christians have been telling us all along: not all Christians are Augustinian. Perhaps an alternative theology can save Christianity from the downfall of Adam and Eve. After all, it is not exactly as if in Genesis 1 through 3 God actually spells out the doctrine of original sin—"scripture does not refer to the Fall, traditionally understood, and nowhere speaks of Adam's sin as a physical inheritance" (Green 2017, 114). It all takes a lot of heavy-duty philosophical and theological reconstruction, often skating quickly over such thin-ice worries as to why (the supposedly originally righteous) Adam and Eve sinned in the first place if they were not already weak or inclined to sin, aided by some nigh-cultish beliefs alien to today's mind frame about the need of blood sacrifices. One rival tradition, as venerable in its way as that of Anselm, is especially powerful in Orthodox circles. The central idea of "Christus Victor" is that we are in bondage and this leads to our sinful nature. Why we are in bondage and to whom is another matter—early views almost all thought it was Satan who had enslaved us and that Christ fought him and triumphed, thus freeing us. The second-century theologian Irenaeus laid it all out. "The Lord therefore ransomed us by his own blood, and gave his life for our life, his flesh for our flesh; and he poured out the Spirit of the Father to bring about the union and fellowship of God and humanity" (McGrath 1995, 176, quoting Irenaeus *Adversus Haereses*). In the last century, very influentially, the Swedish theologian Gustaf Aulén embraced this kind of thinking. "The central theme is the idea of the Atonement as a Divine conflict and a victory; Christ—Christus Victor fights against and triumphs over the evil powers of the world, the 'tyrants' under which mankind is in bondage and suffering, and in Him God reconciles the world to Himself" (Aulén 1931, 4). The Atonement is preserved but there are obviously many questions about this approach, starting with the source of evil. Whereas Irenaeus seems to think of an anti-Christ, Aulén seems to think more in terms of human evildoers. Who needs Satan when you have Hitler? Also, why exactly did the death on the Cross actually conquer evil? The main point for us here is that humans are not tainted because of their own failures. We do wrong and we fail, but because it was shoved onto us and we do not fail inevitably.

Similar thinking can be found in the other non-Augustinian approach, also with long-standing roots. "Incarnational theology" looks upon Jesus as the perfect human and hence a role model for us all. The death on the Cross was the ultimate act and inspiration. "This 'bringer of glad tidings' died as he had lived, as he *taught—not* to 'redeem mankind' but to demonstrate how one ought to live" (Nietzsche 1895, 159). This was the position of Clement of Alexandria. "God himself is love, and for the sake of this love he made himself known." Continuing: "What is the nature and the extent of this love? For each of us he laid down his life, the life which was worth the whole universe, and he requires in return that we should do the same for each other" (McGrath 1995, 177, quoting Clement of Alexandria, *Quis Dives Salvetur*). Peter Abelard followed up on this kind of thinking: "our redemption through the suffering of Christ is that deeper love within us which not only frees us from slavery to sin, but also secures for us the true liberty of the children of God, in order that we might do all things out of love rather than out of fear" (McGrath 1995, 184, quoting Peter Abelard, *Expositio in Epistolam ad Romanos*).

This is the theology of the Jesus of the Beatitudes, not of St. Paul who spent his life worrying about not being worthy. (A recent, sophisticated discussion of biblical evidence for original sin (Green 2017)—one positive in arguing that the idea "*was developed from scriptural warrants*" (115)—jumps right from the Old Testament across to St. Paul.) Probably showing my prejudices, this Jesus-focused theology was very much that of my Quaker childhood, over sixty years ago. It was also, showing that non-conformist though I may have been I did not stand entirely outside the establishment thought of my country of birth, a theology of great appeal to those Anglicans in the late nineteenth century trying to come to terms with the fact of Darwinian evolution. It was a theology that, for people like Aubrey Moore, made evolutionary thinking a positive lynchpin in their Christian commitment, a welcome support not a dreadful challenge. Adam and Eve are irrelevant because we are not into Atonement, blood sacrifices to lift past sins. We are into Incarnation, an exemplar of and inspiration for how we all should behave. A pastor who loves us in our weakness and offers hope and forgiveness.

Here my point is not to endorse or suggest that this option or other non-Augustinian options have no problems. My point is to stress again that we have positions that do not lay our sinful nature on prior sins committed. No one is denying that we are sinful. Of course, we are sinful. Niebuhr knew whereof he spoke when he quipped that "the doctrine of original sin is the only empirically verifiable doctrine of the Christian faith" (Niebuhr 1965, 24). The point is whether it is bound to happen in the way that the Augustinian demands. There is the added question of whether we are now looking at approaches that could take an evolutionary interpretation. Not whether they were formulated with an eye to such an interpretation, but whether as Aubrey Moore thought one could be imposed or integrated. I see no reason why not. Much more positively, the minority Christians we have considered in this book—notably men like Clyde Allee and Bill Thorpe who were professional biologists—give one reason for hope. In non-Augustinian theology, it is quite plausible to suggest that we humans are in a state of becoming rather than being. We are imperfect but Christ's incarnation and death on the cross either makes possible improvement or points the way to such improvement, while encouraging us through being at one with us. It does put responsibility on us. The Sermon on the Mount is absolute. This means we must have total freedom to choose between good and evil. The theology of Immanuel Kant with his propensity toward evil rather than a predisposition. We cannot be innately evil or innately tainted with evil. We do too often choose evil, but we are responsible and cannot put it on the shoulders of a very distant, naughty ancestor. Responsible but not friendless. We are not alone. Jesus is much more man than distant deity. Remember those powerful words of the fourth gospel: "And the Word was made flesh, and dwelt among us, (and we beheld his glory, the glory as of the only begotten of the Father), full of grace and truth" (John

1:14). He is with us because he loves us and he wants to help us. Jesus, in Vermeer's great painting, with his beloved friends Mary and Martha. Not Jesus wracked in agony on the cross, in a Grünewald crucifixion, paying a blood debt to quell the wrath of a vengeful God.

Not that anyone is denying the struggles and pain of Darwinian evolution. Indeed, those in the twentieth century who most openly embraced an evolutionary interpretation of Christianity—Alfred North Whitehead and Pierre Teilhard de Chardin, to take the two most prominent—generally stressed the labor of creation. Thus the Whitehead-influenced, Australian poet Judith Wright:

> Pain, what is it? That which keeps alive
> amoebae doubling from the acid; pain
> that forces flesh to wisdom: hedge of swords
> beside the road from protoplasm to man.

Ironically, this fits very nicely with Darwinism. Richard Dawkins (1983) of all people has argued that the only way in which evolution could produce the design-like features of living beings—adaptations like the hand and the eye—is through natural selection! Be this as it may, and in no sense denying that a huge amount of just war theory and related theorizing still stands, the point is that there are reasons now to break from much that was argued in the past and to look forward to more fruitful venues. This, to the Christian as well as to others, might be a very good thing indeed.

Killer Apes?

We are not necessarily locked into the traditional Christian take on human nature and its implications for understanding war. Turn now to Darwinism and in particular to its discussion of our supposed violent heritage. This is as crucial to the majority Darwinian take on war—and a reason for worrying about its optimism—as original sin is to the majority Christian take on war—and a reason for worrying about its pessimism. A recent book puts firmly the position we have been describing.

> Like all living things, *Homo sapiens* possess an ancient heritage; over the course of many millions of years, the forces of evolution have honed and sculpted our minds and bodies, and this patrimony has an enormous impact on how we live our lives today. The genetic programing bequeathed to us by our ancestors has many constructive, life-affirming aspects. It incites us

> to seek attractive mates, to savor the flavor of nourishing food, to nurture our children, to understand and control the world around us, and even to compose exquisite music and create wonderful works of art. However, our evolutionary legacy also has a much more disturbing face: it moves us to kill our fellow human beings. Violence has followed our species every step of the way in its long journey through time. From the scalped bodies of ancient warriors to the suicide bombers in today's newspaper headlines, history is drenched in human blood. (Livingstone Smith 2007, 8)

The author continues: "Human beings wage war because it is in our nature to do so, and saying that war is just a matter of choice without taking account how our choices grow in the rich soil of human nature is a recipe for confusion. The question that needs answering is, 'What in human nature causes us to choose war?'" Elaborating: "Those who in principle reject an evolutionary account of collective human violence must either deny its existence—which is surely a quixotic move—or else provide an alternative hypothesis. So far, no coherent alternative has been suggested" (26).

These are the words of a philosopher. You can find the same attitude expressed in the world of the empirical scientist. For instance, in a recent collection, USC biological anthropologist Christopher Boehm argues strongly that we see the roots of warfare in the chimpanzee-human analogy. Methodologically, we can look at living groups (clades) of great apes, in order "to see if certain behaviors are shared with the living clade and therefore are very likely—given the parsimony with which evolutionary processes work—to relate to a shared ancestor" (Boehm 2013, 316). On this ground, Boehm concludes that while we can certainly see peacemaking elements, "the conservative, historical starting point is a markedly hierarchical, moderately territorial Ancestral *Pan* [chimpanzee] that was capable of wounding and likely killing other members of the same group" (317). Continuing that "it had no mechanism for peacemaking at the intergroup level." This then translates up into the human level, where the need of hunting abilities combined violent attitudes with cooperation with fellows. "This cooperation, along with the exposure to danger and the explicit objective of killing, would seem to articulate quite easily into 'raiding' or 'warfare' patterns, while an accompanying 'merit system' might have stimulated and prestigiously rewarded both types of behavior" (319). Based on anthropological and archeological evidence, Boehm takes war back a long way in our prehistory. More or less as far as we are able to take it. "Thus, for at least from 45,000 BP to 15,000 BP, or a bit later, it seems likely that in all human societies a combination of chronic male competition over females and recurrent intensive resource competition were stimulating some level of lethal conflict between bands" (326). Little wonder that Boehm concludes that, for humans, war is in some sense natural.

Now, let us bring the same skepticism to Darwinism as we brought to Christianity. The big problem with Christianity is the idea that war is inevitable, or at least predictively certain, because of our sinful nature. Blow this Providence-based picture aside and you can start to rethink things. A big problem with Darwinism is the idea that war is part of our nature, in a sense a good thing, but something we can now conquer. Blow this progress-based picture aside and you can start to rethink things. As we have gone after the Augustinian notion of original sin, let us start by going after the Darwinian notion of the killer ape, something that then leads to thoughts of changing and (perhaps naïve) optimistic hopes of success. This seems a promising strategy if only because, as we have seen, so much of the killer-ape thinking is deeply rooted in Christian thinking about original sin. Remember, we are "Cain's children." It is promising also because it is precisely those for whom Augustinian original sin would not be an influence, consciously or subconsciously—Kropotkin, Allee, Thorpe, Montagu, to name some—who are those over whom the innate killer-ape scenario has no hold.

Chicago physical anthropologist Russell H. Tuttle (2014), in a massive recent overview of what we now know about primate behavior and its relevance to humankind, *Apes and Human Evolution*, shows how, even putting the Christian connections on one side, much of the writing of the past century should be regarded with (at the least) suspicion. Such past writing revealingly often shows more about the concerns of the day, than anything that can be reliably inferred from the evidence. Tuttle argues that after the two great wars we see—as indeed we can attest to having seen in earlier chapters—that people tended to reflect their worries into the conclusions that they draw about human nature. There was a great deal of "man the killer ape" sort of writing predominating. We saw how Tuttle links this to the Vietnam War and related tensions. Then, continues Tuttle, having taken time out for a softer, more feminism-friendly approach, by the end of the century, egged on by the ongoing conflicts around the globe, man the hunter reappeared—"inspired by extensive observations of chimpanzees capturing monkeys and other mammals of modest size and apparently sharing the prey among individuals that were present at the kill site" (10). Referring somewhat sardonically to "people the prejudiced,"

Tuttle concludes that we have had too much ideology chasing too few facts, and certainly too few facts put properly in context.

Going now more on the attack, Tuttle argues that we must not underestimate the importance of the development of attributes (like language ability) that make possible culture and of how these start to make more and more improbable the hypothesis of man the innate killer ape. "It is tempting to view the violent acts of some chimpanzees against conspecifics as analogous to human homicide, infanticide, genocide, or other cruelties toward vulnerable individuals and groups." He continues: "Some consider this evidence of a special close biological linkage between chimpanzees and humans, perhaps with roots as far back as the Early Pliocene, and they support such scenarios by placing demonic males predominant in hominid phylogeny" (592–593). Tuttle will have none of it. He thinks such language just plain silly and notes that there simply is no paleoanthropological evidence backing such views. The biology is against it. "Indeed, given the lack of fearsome teeth and other morphological or technological means to kill one another, the arguable probability is that they lived consistently in groups with sizable cohorts of adult males under an omnipresent possibility that they might be confronted by carnivoran predators, which were more formidable in number and variety than modern ones" (593). The case is powerful that simplistic scenarios of the killer ape, with blood dripping from its fangs—the monster of the anti-German posters of the First World War—just don't have serious empirical or theoretical backing. "Because there were fewer hominids in the Pliocene and Pleistocene, group migration away from agonistic groups would seem to have been an available option. Moreover, developing mutually beneficial relationships with them probably would be a better option for one or both parties, instead of having to plan and execute attacks in lieu of foraging and sustaining bonds within one's own group" (594).

Others sing a similar theme. Some just don't much care for the use of apes as guides to humans. Notre Dame anthropologist Agustín Fuentes (2013) notes that while "humans and the two *Pan* species may share some similarities, . . . behaviorally they are quite divergent in most areas. Simply put, we must stop envisioning chimpanzees and bonobos [pygmy chimps] as a model (behaviorally, morphologically, ecologically) for the RCA [recent common ancestor]" (85). Others point to the fact that the bonobos simply are not aggressive in the way of the regular chimpanzee. In fact, their social life is not unlike that of undergraduates at Canadian universities in the 1960s—sex, more sex, and yet more sex. D. H. Lawrence's dream world. Frans de Waal, he who finally told us about the violence between male chimpanzees, is categorical on this subject. "Bonobos have never been reported to kill each other, neither in captivity nor in the field. They sometimes mingle across territorial borders, and are known as 'make love—not war' primates for solving power issues through sexual activity" (de Waal 2013, xii). It gives a whole new meaning to the notion of "soldiers erect."

Turning to our species, Pinker's claims about our violent past are subjected to withering analysis and scorn. He biased the picking of data, he misinterpreted the evidence, and he drew false and misleading conclusions. "By considering the *total* archaeological record of prehistoric populations of Europe and the Near East up to the Bronze Age, evidence clearly demonstrates that war began sporadically out of warless condition, and can be seen, in varying trajectories in different areas, to develop over times as societies become larger, more sedentary, more complex, more bounded, more hierarchical, and in one critically important region, impacted by an expanding state" (Ferguson 2013, 116). Note that these critics are not breaking totally with past thinking. No one is denying that people have been warlike. Michigan anthropologist Raymond C. Kelly (2000) makes it clear that he does not think war is an inevitable effusion of human nature. "Warfare is not an endemic condition of human existence but an episodic feature of human history (and prehistory) observed at certain times and places but not others" (75). Not inevitable, but war does happen and the interesting and important question is what makes for war or non-war. No one is denying that the kinds of cultural factors that have been raised are very important, if not crucial. The move to agriculture for instance with the new existence of desirable fertile lands (that you cannot pick up and move with to avoid conflict), of stockpiles of food, of increased population number, and perceived need of more territory—shades of Germany in both the First and Second World—and the growth of sophisticated political systems as people live sedentary lives and increase in numbers.

> An important milestone in looking at the origins of warfare in humans is 8000 BC, as it stands at the very end of the Mesolithic and the beginning of the Neolithic periods. It also marks major changes in the trajectory of human history as humankind was reaching the upper demographic limits of sustainable hunting and gathering around the world. People were in the throes of the transition from a hunting-and-gathering, nomadic lifestyle to an agricultural and settled lifestyle. (Haas and Piscitelli 2013, 178)

War came in afterwards. Looking back one hundred thousand to two hundred thousand years, there is simply no evidence of war in earlier times. This is true, notwithstanding expected debates over causes and details, for instance the extent that capturing women has been a significant factor in bringing on war (Fry 2013, 20). No one is denying that Pinker is quite possibly right and that violence has declined over the centuries, although how one treats the twentieth century might be a matter of debate. The question is whether we were and always have been warlike, as Pinker and so many others claim. That is disputed. "Contra Pinker, the incidence of warfare and violent mayhem over the last 10 millennia actually follows an n-shaped curve, rather than merely the steep drop-off in recent times that Pinker highlights" (15). Continuing, "the archaeological record is clear and unambiguous: war developed, despots arose, violence proliferated, slavery flourished, and the social position of women deteriorated." But it is recent. At most ten thousand years ago and probably quite a bit less. Pinker's Pacification trend got it precisely backward. War came because of the move to agriculture, not departed because of the move to agriculture.

There was incidentally a remarkably prescient anticipation of this kind of thinking by the Dutch-Anglo ethologist Niko Tinbergen in his professorial inaugural lecture at Oxford in 1968. He too questioned the assumption that we have an innate aggressive streak and he too pondered the effects of agriculture and population growth.

> Agricultural and technical know-how have enabled us to grow food and to exploit other natural resources to such an extent that we can still feed (though only just) the enormous numbers of human beings on our crowded planet. The result is that we now live at a far higher density than that in which genetic evolution has molded our species. This, together with long-distance communication, leads to far more frequent, in fact to continuous, intergroup contacts, and so to continuous external provocation of aggression. (Tinbergen 1968, 1415)

Add in the ability to make powerful weapons and the like, and you have a recipe for disaster. Lorenz was very upset by this lecture, thinking that Tinbergen had quite betrayed the ethological movement (Burkhardt 2005, 441).

We have seen all along Darwinians hopeful about the end of war. I want to avoid the Scylla of likewise falling into false optimism, while steering clear of the Charybdis of being too careful to say anything at all, at least anything of interest and importance. No one is denying that we are what we are because of natural selection. The question is whether this struggle always and in a widespread manner involved warfare. The point I take from this section is that the biggest blocking point in the Darwinian account of war—naivety in thinking that we can now reverse the course of millions of years of evolution, not to mention the counterintuitive belief that early warfare must have been good—may not be quite as big as one fears. It could be that naturally we are more peaceful than we give credit for, and that much strife is truly more cultural and hence at least in theory eliminable—"*neither the observable facts nor the application of evolutionary principles supports the notion that war is an evolutionary adaptation*" (Fry 2007, 167, his italics). Darwinism—as understood in the terms of this book—seems as open to rethinking as does Christianity. In part, probably for the same reasons, given the role that a sometime Bishop of Hippo plays in these discussions. In the light of philosopher David Livingstone Smith's enthusiastic endorsement of the killer-ape picture of humankind, I am glad to say philosopher A. C. Grayling (2017) agrees with me.

After War?

We don't have to be held forever in the iron grip of the past. Perhaps—probably—there will always be tensions between Christians and Darwinians. They are looking for different answers. However, the time has come to speak more directly to each other, to recognize that both are dealing with the same problems, and that there is not necessarily an iron barrier that can never be breached. At bottom, we have two visions of human nature but they are not quite as distinct as partisans (on both sides) assume. Neither—on their own terms—need be quite so committed to the view that war has been always an essential element of human nature and development, and that hence (pessimism) we can never get rid of it or (optimism) that we can simply switch practices and avoid it. Neither in theory nor in practice is war a necessary part of our story. The belief that it is comes in major part because of a massive circularity. On theological grounds, Augustinians think we are sinful. This gets picked up by Darwinians and naturalized into the killer-ape hypothesis. And then it is fed back into theology by neo-Augustinians adrift in a world without Adam and Eve.

I am absolutely not saying that what Christians and Darwinians have produced thus far on war is worthless. What I am saying is that it is dangerously incomplete and distorted. Horrendously and cripplingly circular. Ask the question. If what has been produced on war is all that good, why isn't it more effective? Time for new ideas and findings. It is surely open to the Christian, with the aid of the Good God, to hope that we can start the move forward. Likewise, it is surely open to the Darwinian, with the aid of the professional theory, to hope that we can start the move forward. In both cases, we have to think and rethink what we mean by human nature and what this thinking and rethinking means for our understanding of humans and warfare. Is it something in the soul or in the genes, as it were, and cannot be eradicated, can at most be controlled? Or, is it something that is not inevitable and that, given what there is within human nature—soul or genes—we can think seriously about moving on? I am much aware of the underlying worry in discussions like these that truly we are pawns of history, now tossing us one way and now the other. Caught in a never-ending cycle. In our case, the shuttlecock in the game of badminton played by the followers of Thomas Hobbes and the followers of Jean-Jacques Rousseau. For a while we have been in the hold of the violent man always pushing and fighting, the killer ape (for instance, Keeley 1996). Fashions change, and now we have moved to the man of amity, the peaceful savage—"a careful reexamination of the actual evidence will lead us to the conclusion that humans are not warlike by nature" (Fry 2007, 2; and see also Fry 2013 and many of the contributors to his volume). Only for a while, however, and then we shall go back to blood-dripping fangs.

This is a serious concern. Tuttle (2014) has flagged us to the extent to which these discussions reflect the trendy ideology of the day. Yet, not entirely. Take the analogy of the debate in biology over form and function (Russell 1916; Ruse 2017b). These notions have played out seemingly endlessly in evolutionary circles, now going one way and now the other. There has been advance. The evidence is simply overwhelming that to understand the living world, in some sense function must be prior. Adaptation rules! This does not mean that form is gone, or that homologies are now judged unimportant. There will surely always be people like Stephen Jay Gould who think that form should be given a bigger role in our thinking. There will always be debate and probably cycles. To borrow from Spencer, however, you will never get back to where you were. There will be advance as the facts come in and the models get more sophisticated. A kind of dynamic equilibrium.

In the specific case of war, it is not naïve optimism to note that in the past half century there has been a quantum leap in sophistication by theologians and philosophers and others thinking about the nature of the phenomenon, its scope, and its inevitability. Likewise, there has been a quantum leap in the amount of data and powerful analyses of what we can learn about the prehistory of our species, and the extent to which war is, has been, always was part of our nature and our societies. All of Raymond Dart's worrying about humans as killers, apparently, is based on a misreading of the fossil evidence. Brain (1983) makes a strong case that the signs of violence—scars on the skulls, and so forth—are better interpreted as humans hunted (by carnivores) than humans hunters (of fellow humans). Or take the celebrated case of the Yanomamö Indians. Biological anthropologist Douglas P. Fry (2007) has offered a devastating critique of Chagnon's analysis, showing that he commits just about every statistical sin in the book, starting with the fact that he fails to control properly for age. On the evidence offered, it could simply be that older men have more children. Probably true but not quite what was implied or needed. Whether or not the study really holds up, whether or not the criticisms are really well taken, discussion advances meaningfully. Some change is, if not progress, reason to hope for progress. Pinker for one, for all that he has his critics, would agree with this, as would Christians who have no Augustinian hang-ups. That eighteenth-century child of Pietism, author of the essay on "Perpetual Peace," was one. Move on, but always as Christians and as Darwinians, with an eye to the past. Christians have much to learn from Darwinism and, in looking at the history of Christian theorizing about war, there are deep insights that should help Darwinians to focus on their interests and understandings. War is a horrible thing. We should work together toward a bigger picture.

The reader might complain that, under the guise of a disinterested account of the differences between Christians and Darwinians on war, I am implicitly preaching what I call Darwinism, that it will be a brighter day tomorrow. What would one expect from a non-believing Darwinian, raised a Quaker? Not so! I distance myself from claims about biological progress. I am a Humean, thinking value is imputed not found. I am not a pacifist. Hitler had to be stopped. I am far from optimistic about the long-term prospects of humankind. The fate of all species is extinction. My Inner Darwinian tells me that we must keep trying. Not to strive for social progress is to guarantee that such progress will not occur. The meek do not inherit the earth.

BIBLIOGRAPHY

Abbott, L. 1918. *The Twentieth Century Crusade*. New York: Macmillan.

Adams, R. M. 1999. Original sin: A study in the interaction of philosophy and theology. *The Question of Christian Philosophy Today.* Editor F. J. Ambrosio, 80–110. New York: Fordham University Press.

Agassiz, L. 1859. *Essay on Classification*. London: Longman, Brown, Green, Longmans, and Roberts and Trubner.

Allee, W. C. 1927. Animal aggregations. *Quarterly Review of Biology* 2: 367–398.

———. 1938. *The Social Life of Animals*. London: Heinemann.

———. 1943. Where angels fear to tread: A contribution from general sociology to human ethics. *Science* 97: 517–525.

Allen, G. E. 1978. *Life Science in the Twentieth Century*. Cambridge: Cambridge University Press.

Anon. [D. T. Anstead] 1860a. Natural selection. *All the Year Round* 3 (63): 293–299.

———. 1860b. Species. *All the Year Round* 3 (58): 174–178.

Anon. November 11, 1932. Mr Baldwin on aerial warfare—a fear for the future. *The Times*, col. B p. 7.

Anonymous, editor. 1860. *Essays and Reviews*. London: Longman, Green, Longman, and Roberts.

Anscombe, G. E. M. [1957] 1981. Mr Truman's degree. *Ethics, Religion and Politics: Collected Philosophical Papers, Volume III.* G. E. M. Anscombe, 62–71. Oxford: Blackwell.

———. 1981. *Ethics, Religion and Politics: Collected Philosophical Papers, Volume III.* Oxford: Blackwell.

Anscombe, G. E. M., and N. Daniel. [1939] 1981. The justice of the present war examined. *Ethics, Religion and Politics: Collected Philosophical Papers, Volume III.* G. E. M. Anscombe, 72–81. Oxford: Blackwell.

Ardrey, R. 1961. *African Genesis: A Personal Investigation into the Animal Origins and Nature of Man*. New York: Atheneum.

Atkinson, R. 2002. *An Army at Dawn: The War in North Africa*, 1942–1943. New York: Henry Holt.

———. 2007. *The Day of Battle: The War in Sicily and Italy*, 1943–1944. New York: Henry Holt.

———. 2013. *The Guns at Last Light: The War in Western Europe*, 1944–1945. New York: Henry Holt.

Augustine. [413–426] 1998. *The City of God against the Pagans*. Editor and translator R. W. Dyson. Cambridge: Cambridge University Press.

———. 1886. *On the Trinity*. Translator A. W. Hadden. New York: Philip Schaff.

Aulén, G. [1931] 2003. *Christus Victor: An Historical Study of the Three Main Types of the Idea of the Atonement*. Eugene, OR: Wipf & Stock.

Ayala, F. J. 2009. Molecular evolution. *Evolution: The First Four Billion Years*. Editors M. Ruse and J. Travis, 132–151. Cambridge, MA: Harvard University Press.

Bagehot, W. 1868. *Physics and Politics: Or Thoughts on the Application of the Principles of "Natural Selection" and "Inheritance" to Political Society*. London: Henry S. King.

Baker, J. R. 1978. *Julian Huxley, Scientist and World Citizen, 1887–1975*. Paris: UNESCO.

Bang, J. P. 1917. *Hurrah and Hallelujah: The Teaching of Germany's Poets, Prophets, Professors and Preachers*. New York: George H. Doran.

Barash, D. P., and J. E. Lipton. 1985. *The Caveman and the Bomb: Human Nature, Evolution, and Nuclear War*. New York: McGraw-Hill.

Barrett, C. 2014. *Subversive Peacemakers: War Resistance 1914–1918. An Anglican Perspective*. Cambridge: Lutterworth Press.

Barrett, P. H., P. J. Gautrey, S. Herbert, D. Kohn, and S. Smith, editors. 1987. *Charles Darwin's Notebooks, 1836–1844*. Ithaca, NY: Cornell University Press.

Barth, K. 1933. *The Epistle to the Romans*. Oxford: Oxford University Press.

———. [1934] 1962. The theological declaration of Barmen. *The Church's Confessions under Hitler*. Editor A. C. Cochrane, 237–242. Philadelphia: Westminster Press.

———. 1941. *This Christian Cause (A Letter to Great Britain from Switzerland)*. New York: Macmillan.

———. [1951] 1961. *Church Dogmatics, III, 4. The Doctrine of Creation*. Edinburgh: T. and T. Clark.

Baynes, T. S. 1873. Darwin on expression. *Edinburgh Review* 137: 492–508.

Bell, S. 2012. The Church and the First World War. *God and War: The Church of England and Armed Conflict in the Twentieth Century*. Editors S. G. Parker and T. Lawson, 33–60. London: Routledge.

Benedict XV. 1915. *Apostolic exhortation: To the peoples now at war and to their rulers*. *Ecclesiastical Review* liii: 434–437.

Bergson, H. 1911. *Creative Evolution*. New York: Holt.

Biggar, N. 2013. *In Defence of War*. Oxford: Oxford University Press.

Boehm, C. 2013. The biocultural evolution of conflict resolution between groups. *War, Peace, and Human Nature: The Convergence of Evolutionary and Cultural Views*. Editor D. P. Fry, 315–340. Oxford: Oxford University Press.

Bourne, R. S. [1917] 1999. Twilight of idols. *War and the Intellectuals: Collected Essays, 1915–1919*. R. S. Bourne, 53–64. Indianapolis, IN: Hackett.

Bowler, P. J. 1983. *The Eclipse of Darwinism: Anti-Darwinism Evolution Theories in the Decades around 1900*. Baltimore: Johns Hopkins University Press.

———. 1984. *Evolution: The History of an Idea*. Berkeley: University of California Press.

———. 1988. *The Non-Darwinian Revolution: Reinterpreting a Historical Myth*. Baltimore: Johns Hopkins University Press.

———. 1989. *The Mendelian Revolution: The Emergence of Hereditarian Concepts in Modern Science and Society*. London: The Athlone Press.

———. 2013. *Darwin Deleted: Imagining a World without Darwin*. Chicago: University of Chicago Press.

Brain, C. K. 1983. *The Hunters or the Hunted?: An Introduction to African Cave Taphonomy*. Chicago: University of Chicago Press.

Brodie, R. J. 1845. *The Secret Companion*. London: R. J. Brodie.

Browne, J. 1995. *Charles Darwin: Voyaging. Volume I of a Biography*. New York: Knopf.

———. 2002. *Charles Darwin: The Power of Place. Volume II of a Biography*. New York: Knopf.

Bryan, W. J. 1919. The most tragic biological blunder ever made. *Current Opinion* 66: 303–304.

———. 1922. The origin of man. *In His Image*, 86–135. New York and Chicago: Fleming H. Revell.

Buchan, J. [1916] 1992. *Greenmantle. The Complete Richard Hannay*. London: Penguin.

Bud, R. 2013. Life, DNA and the model. *British Journal for the History of Science* 46: 311–334.

Burkhardt, R. W. 2005. *Patterns of Behavior: Konrad Lorenz, Niko Tinbergen, and the Founding of Ethology*. Chicago: University of Chicago Press.

Burroughs, E. R. [1912] 1914. *Tarzan of the Apes*. Chicago: McClurg.

Calvin, J. [1536] 1960. *Institutes of the Christian Religion*. Grand Rapids, MI: Eerdmans.

Carpenter, H., and C. Tolkien, editors. 2000. *The Letters of J. R. R. Tolkien*. New York: Mariner Books.

Chalmers Mitchell, P. 1915. *Evolution and the War*. London: John Murray.

Chambers, R. 1844. *Vestiges of the Natural History of Creation*. London: Churchill.

Chandler, A. 2016. *The Church and Humanity: The Life and Work of George Bell, 1883–1958*. London: Routledge.

Clark, C. 2006. *The Iron Kingdom: The Rise and Downfall of Prussia, 1600–1947*. Cambridge, MA: Harvard University Press.

———. 2014. *The Sleepwalkers: How Europe Went to War in 1914*. New York: Harper.

Clark, R. W. 1968. *J. B. S.: The Life and Work of J. B. S. Haldane*. London: Hodder and Stoughton.

Comstock, J. H. 1893. Evolution and taxonomy. *The Wilder Quarter Century Book*, 37–114. Ithaca, NY: Comstock.

Conklin, E. G. 1921. *The Direction of Human Evolution*. London: Oxford University Press.

Conway, J. S. [1968] 1997. *The Nazi Persecution of the Churches, 1933–1945*. Vancouver, Canada: Regent College Publishing.

Conway Morris, S. 2003. *Life's Solution: Inevitable Humans in a Lonely Universe*. Cambridge: Cambridge University Press.

Coyne, J. A. 2015. *Faith Versus Fact: Why Science and Religion Are Incompatible*. New York: Viking.

Cunningham, S. 1996. *Philosophy and the Darwinian Legacy*. Rochester, NY: University of Rochester Press.

Dart, R. 1953. The predatory transition from ape to man. *International Anthropological and Linguistic Review* 1 (4): 201–217.
Darwin, C. 1859. *On the Origin of Species by Means of Natural Selection, or the Preservation of Favoured Races in the Struggle for Life*. London: John Murray.
———. 1861. *Origin of Species (Third Edition)*. London: John Murray.
———. 1868. *The Variation of Animals and Plants under Domestication*. London: John Murray.
———. 1871. *The Descent of Man, and Selection in Relation to Sex*. London: John Murray.
———. 1874. *The Descent of Man (Second Edition)*. London: John Murray.
———. 1958. *The Autobiography of Charles Darwin 1809–1882. With the Original Omissions Restored. Edited and with Appendix and Notes by his Grand-daughter Nora Barlow*. London: Collins.
———. 1985–. *The Correspondence of Charles Darwin*. Cambridge: Cambridge University Press.
Darwin, E. 1803. *The Temple of Nature*. London: J. Johnson.
Davidson, R. T. 1919. *The Testing of a Nation*. London: Macmillan.
Dawkins, R. 1976. *The Selfish Gene*. Oxford: Oxford University Press.
———. 1983. Universal Darwinism. *Evolution from Molecules to Men*. Editor D. S. Bendall, 403–425. Cambridge: Cambridge University Press.
———. 1986. *The Blind Watchmaker*. New York: W. W. Norton.
———. 1995. *A River Out of Eden*. New York: Basic Books.
———. 1997. Human chauvinism: Review of *Full House* by Stephen Jay Gould. *Evolution* 51 (3): 1015–1020.
———. 2006. *The God Delusion*. New York: Houghton, Mifflin, Harcourt.
Dawkins, R., and J. R. Krebs. 1979. Arms races between and within species. *Proceedings of the Royal Society of London, B* 205: 489–511.
De Waal, F. 1982. *Chimpanzee Politics: Power and Sex among Apes*. London: Cape.
———. 2007. *Chimpanzee Politics: Power and Sex among Apes: 25th Anniversary Edition*. Baltimore: Johns Hopkins University Press.
———. 2014. Foreword. *War, Peace, and Human Nature*. Editor D. P. Fry, xi–xiv. Oxford: Oxford University Press.
Deane-Drummond, C. 2017. In Adam all die: Questions at the boundary of niche construction, community evolution, and original sin. *Evolution and the Fall*. Editors W. T. Cavanaugh and J. K. A. Smith, 23–47. Grand Rapids, MI: Eerdmans.
Dembski, W. A., and M. Ruse, editors. 2004. *Debating Design: Darwin to DNA*. Cambridge: Cambridge University Press.
Desmond, A. 1997. *Huxley: From Devil's Disciple to Evolution's High Priest*. New York: Basic Books.
Desmond, A., and J. Moore. 2009. *Darwin's Sacred Cause: How a Hatred of Slavery Shaped Darwin's Views on Human Evolution*. New York: Houghton Mifflin Harcourt.
Dixon, E. S. 1862. A vision of animal existences. *Cornhill Magazine* 5 (27): 311–318.
Dobzhansky, T. 1937. *Genetics and the Origin of Species*. New York: Columbia University Press.
Domning, D. P. 2001. Evolution, evil and original sin. *America* 185: 14–21.

Eibl-Eibesfeldt, I. 1979. *The Biology of Peace and War: Men, Animals, and Aggression.* New York: Viking.

Evans, R. J. 2003. *The Coming of the Third Reich.* New York: Penguin.

Falk, D. 2012. *The Fossil Chronicles: How Two Controversial Discoveries Changed Our View of Human Evolution.* Berkeley: University of California Press.

Farrar, F. W. 1858. *Eric or, Little by Little.* Edinburgh: Adam and Charles Black.

Ferguson, R. B. 2013. Pinker's list: Exaggerating prehistory war mortality. *War, Peace, Human Nature: The Convergence of Evolutionary and Cultural Views.* Editor D. P. Fry, 112–131. Oxford: Oxford University Press.

Fichte, J. G. [1808] 1922. *Addresses to the German Nation.* Chicago: Open Court.

Fisher, D. 2014. Just war and the First World War: Where was the just war tradition when it was needed? (or how moral theology could have saved the world). *Crucible, the Journal of Christian Social Ethics*, April–June, 27.

Fisher, R. A. 1916. Review of Kellogg, V. L. Military selection and race deterioration. *Eugenics Review* 8: 264–265.

———. 1930. *The Genetical Theory of Natural Selection.* Oxford: Oxford University Press.

———. 1947. The renaissance of Darwinism. *Listener* 37: 1001.

Fodor, J., and M. Piattelli-Palmarini. 2010. *What Darwin Got Wrong.* New York: Farrar, Straus, and Giroux.

Ford, J. C. 1944. The morality of obliteration bombing. *Theological Studies* 5: 261–309.

Fosdick, H. E. 1917. *The Challenge of the Present Crisis.* New York: Association Press.

———. 1922. Shall the fundamentalists win? *Christian Work,* 102 (June 10): 716–722.

———. 1925. *A Christian Conscience about War: Sermon delivered to League of Nations Assembly Service, Cathedral, Geneva, Sept. 13, 1925.* New York: NP.

———. 1934. *The Secret of Victorious Living: Sermons on Christianity Today.* New York and London: Harper and Brothers.

Fry, D. P. 2007. *Beyond War: The Human Potential for Peace.* Oxford: Oxford University Press.

———. 2013. War, peace, and human nature: The challenge of achieving scientific objectivity. *War, Peace, and Human Nature: The Convergence of Evolutionary and Cultural Views.* Editor D. P. Fry, 1–21. Oxford: Oxford University Press.

Fuentes, A. 2013. Cooperation, conflict, and niche construction in the genus *Homo. War, Peace, and Human Nature.* Editor D. P. Fry, 78–94. Oxford: Oxford University Press.

Fussell, P. 1975. *The Great War and Modern Memory.* New York: Oxford University Press.

Gasman, D. 1998. *Haeckel's Monism and the Birth of Fascist Ideology.* Frankfurt: Peter Lang.

German Christian Churches. 1908. *Peace and the Churches. Souvenir Volume of the Visit to England of Representatives of the German Christian Churches. May 26th to June 3rd, 1908. Including the Visit to Scotland, June 3rd to 7th, 1908; Der Friede und die Kirchen.* London: Cassell.

Gibson, A. 2013. Edward O. Wilson and the organicist tradition. *Journal of the History of Biology* 46: 599–630.

Gilkey, L. B. 2001. *On Niebuhr: A Theological Study.* Chicago: University of Chicago Press.

Gissing, G. 1976. *New Grub Street.* London: Penguin.

Gluckman, P., A. Beedle, and M. Hanson. 2009. *Principles of Evolutionary Medicine*. Oxford: Oxford University Press.

Goff, S. 2015. *Borderline: Reflections on War, Sex, and Church*. Eugene, OR: Cascade.

Goodall, J. 1986. *The Chimpanzees of Gombe: Patterns of Behavior*. Cambridge, MA: Belknap.

Gore, C. 1889. *Lux Mundi*. London: John Murray.

Gould, S. J. 1977. *Ever Since Darwin*. New York: W. W. Norton.

———. 1985. *The Flamingo's Smile: Reflections in Natural History*. New York: W. W. Norton.

———. 1989. *Wonderful Life: The Burgess Shale and the Nature of History*. New York: W. W. Norton.

———. 1999. *Rocks of Ages: Science and Religion in the Fullness of Life*. New York: Ballantine.

Gould, S. J., and R. C. Lewontin. 1979. The spandrels of San Marco and the Panglossian paradigm: A critique of the adaptationist programme. *Proceedings of the Royal Society of London, Series B: Biological Sciences* 205: 581–598.

Gray, A. [1860] 1876. [Review of] The Origin of Species by Means of Natural Selection, American Journal of Arts and Sciences. In *Darwiniana*. A. Gray, 7–50. New York: D. Appleton.

Gray, A. 1876. *Darwiniana*. New York: D. Appleton.

Grayling, A. C. 2017. *War: An Enquiry*. New Haven, CT: Yale University Press.

Green, J. B. 2017. "Adam, what have you done?" New Testament voices on the origin of sin. *Evolution and the Fall*. Editors W. T. Cavenaugh and J. K. A. Smith, 98–116. Grand Rapids, MI: Eerdmans.

Greg, W. R. 1868. On the failure of "natural selection" in the case of man. *Fraser's Magazine* September: 353–362.

Haas, J., and M. Piscitelli. 2013. The prehistory of warfare: misled by ethnography. *War, Peace, and Human Nature: The Convergence of Evolutionary and Cultural Views*. Editor D. P. Fry, 168–190. Oxford: Oxford University Press.

Haldane, J. B. S. 1925. *Callinicus: A Defence of Chemical Warfare*. New York: Dutton.

———. 1934. *Human Biology and Politics (The Norman Lockyer Lecture)*. London: The British Science Guild.

———. 1938. *A. R. P.* London: Victor Gollancz.

Hamilton, W. D. 1964. The genetical evolution of social behaviour. *Journal of Theoretical Biology* 7: 1–52.

Hardy, T. 1994. *Collected Poems*. Ware, Hertfordshire: Wordsworth Poetry Library.

Hastings, M. 2013. *Catastrophe 1914: Europe Goes to War*. New York: Knopf.

Hauerwas, S. [1984] 2001. Should war be eliminated? A thought experiment. *The Hauerwas Reader*. S. Hauerwas, 392–425. Durham, NC: Duke University Press.

———. [1988] 2001. On being a church capable of addressing a world at war: A pacifist response to the United Methodist Bishops' pastoral *In Defense of Creation*. *The Hauerwas Reader*. S. Hauerwas, 426–458. Durham, NC: Duke University Press.

Hegel, G. W. F. [1821] 1991. *Elements of the Philosophy of Right*. Editor A. Wood. Cambridge: University of Cambridge Press.

Hennessy, P, 2001. *The Prime Minister: The Office and Its Holders since 1945*. New York: Palgrave.

Herschel, J. F. W. 1830. *Preliminary Discourse on the Study of Natural Philosophy.* London: Longman, Rees, Orme, Brown, Green, and Longman.
Hillis, N. D. 1918. *The Blot on the Kaiser's 'Scutcheon.* New York: Fleming H. Revell.
Hitler, A. 1925. *Mein Kampf.* London: Secker and Warburg.
Hogben, L. 1998. *Scientific Humanist: An Unauthorised Biography.* Woodbridge, Suffolk: Merlin Press.
Holmes, A. F., editor. 2005. *War and Christian Ethics, Second Edition.* Grand Rapids, MI: Baker.
Holmes, R. L. 1989. *On War and Morality.* Princeton, NJ: Princeton University Press.
Hoover, A. J. 1989. *God, Germany, and Britain in the Great War: A Study in Clerical Nationalism.* New York: Praeger.
Hume, D. 1978. *A Treatise of Human Nature.* Oxford: Oxford University Press.
Huxley, J. S. 1912. *The Individual in the Animal Kingdom.* Cambridge: Cambridge University Press.
———. 1927. *Religion without Revelation.* London: Ernest Benn.
———. 1942. *Evolution: The Modern Synthesis.* London: Allen and Unwin.
———. 1943. *TVA: Adventure in Planning.* London: Scientific Book Club.
———. 1948. *UNESCO: Its Purpose and Its Philosophy.* Washington, DC: Public Affairs Press.
Huxley, L. 1900. *The Life and Letters of Thomas Henry Huxley.* London: Macmillan.
Huxley, T. H. [1860] 1884. The Origin of Species. In *Collected Essays: Darwiniana,* 22–79. London: Macmillan.
———. 1879. *Hume.* London: Macmillan.
———. 1893. Evolution and ethics. In *Evolution and Ethics,* 46–116. London: Macmillan.
James, W. [1890] 1952. *Principles of Psychology.* Chicago: Encyclopedia Britannica.
———. 1904. *What Pragmatism Means. What Is Pragmatism?* New York: Library of America.
———. [1910] 1982. The moral equivalent of war. *The Works of William James: Essays in Religion and Morality.* Editors F. Burkhardt and F. Bowers, 162–173. Cambridge, MA: Harvard University Press.
Jenkins, P. 2014. *The Great and Holy War: How World War I Became a Religious Crusade.* New York: HarperOne.
Johanson, D., and M. Edey. 1981. *Lucy: The Beginnings of Humankind.* New York: Simon and Schuster.
Johnson, P. E. 1995. *Reason in the Balance: The Case against Naturalism in Science, Law and Education.* Downers Grove, IL: InterVarsity Press.
Jordan, D. S. 1907. *Human Harvest: A Study of the Decay of Races through the Survival of the Unfit.* Boston: American Unitarian Association.
———. 1913. The eugenics of war. *Eugenics Review* 5: 197–213.
Kalikow, T. J. 1983. Konrad Lorenz's ethological theory: Explanation and ideology, 1938–1943. *Journal of the History of Biology* 16: 39–73.
Kant, I. [1785] 1790. The metaphysical elements of the theory of right. (The first part of the *Metaphysics of Morals.*). *Kant: Political Writings.* 131–175. Cambridge: Cambridge University Press.
———. [1788] 1898. *Critique of Practical Reason.* Translator T. K. Abbott. London: Longmans, Green.

———. [1793] 1998. *Religion within the Bounds of Mere Reason*. Editors A. Wood and G. di Giovanni. Cambridge: Cambridge University Press.

———. [1795] 1970. Perpetual peace: a philosophical sketch. *Kant: Political Writings*, 93–130. Cambridge: Cambridge University Press.

Keegan, J. 1999. *The First World War*. New York: Knopf.

Keeley, L. H. 1996. *War before Civilization: The Myth of the Peaceful Savage*. Oxford: Oxford University Press.

Keith, A. 1931. *The Place of Prejudice in Modern Civilization: Prejudice and Politics*. London: Williams and Norgate.

———. 1947. *Evolution and Ethics*. New York: G. P. Putnam's Sons.

———. 1950. *An Autobiography*. New York: Philosophical Library.

Kellogg, V. L. 1905. *Darwinism Today*. New York: Henry Holt.

———. 1912. *Beyond War: A Chapter in the Natural History of Man*. New York: Henry Holt.

———. 1916. *Military Selection and Race Deterioration*. Oxford: Clarendon Press.

———. 1917. *Headquarters Nights: A Record of Conversations and Experiences at the Headquarters of the German Army in France and Belgium*. Boston: Atlantic Monthly Press.

———. 1919. *Germany in the War and After*. New York: Macmillan.

Kelly, R. C. 2000. *Warless Societies and the Origin of War*. Ann Arbor: University of Michigan Press.

Kelsay, J. 2007. *Arguing the Just War in Islam*. Cambridge, MA: Harvard University Press.

Kershaw, I. 1999. *Hitler 1889–1936: Hubris*. New York: W. W. Norton.

———. 2015. *To Hell and Back: Europe 1914–1949 (The Penguin History of Europe)*. New York: Viking.

Kevles, D. J. 1985. *In the Name of Eugenics: Genetics and the Uses of Human Heredity*. New York: Knopf.

Kimler, W., and M. Ruse. 2013. Mimicry and camouflage. *The Cambridge Encyclopedia of Darwin and Evolutionary Thought*. Editor M. Ruse, 139–145. Cambridge: Cambridge University Press.

Kimura, M. 1983. *The Neutral Theory of Molecular Evolution*. Cambridge: Cambridge University Press.

Kirby, W., and W. Spence. 1815–1828. *An Introduction to Entomology: or Elements of the Natural History of Insects*. London: Longman, Hurst, Reece, Orme, and Brown.

Krischel, M. 2010. Perceived hereditary effect of World War I: A study of the positions of Friedrich von Bernhardi and Vernon Kellogg. *Medicine Studies* 2: 139–150.

Kropotkin, P. 1902. *Mutual Aid: A Factor in Evolution*. Boston: Extending Horizons Books.

Larson, E. J. 1997. *Summer for the Gods: The Scopes Trial and America's Continuing Debate over Science and Religion*. New York: Basic Books.

Levins, R., and R. C. Lewontin. 1985. *The Dialectical Biologist*. Cambridge, MA: Harvard University Press.

Lewis, C. S. [1945] 1996. *That Hideous Strength*. New York: Simon and Schuster.

———. 1976. Why I am not a pacifist. *The Weight of Glory*, 64–90. New York: HarperCollins.

Lewontin, R. C. 1974. *The Genetic Basis of Evolutionary Change*. New York: Columbia University Press.

———. 2011. It's even less in your genes. *New York Review of Books*, no. May 26.
———, J. A. Moore, W. B. Provine, and B. Wallace, editors. 1981. *Dobzhansky's Genetics of Natural Populations I–XLIII.* New York: Columbia University Press.
Lightman, B. 2010. Darwin and the popularization of evolution. *Notes and Records of the Royal Society* 64: 5–24.
Lindsey, H., and C. C. Carlson. 1970. *The Late Great Planet Earth.* Grand Rapids, MI.: Zondervan.
Livingstone Smith, D. 2007. *The Most Dangerous Animal: Human Nature and the Origins of War.* New York: St. Martin's Press.
Loeb, J. 1917. Biology and war. *Science* 45: 73–76.
London, J. [1903] 1990. *The Call of the Wild.* New York: Dover.
Lorenz, K. 1966. *On Aggression.* London: Methuen.
Lurie, E. 1960. *Louis Agassiz: A Life in Science.* Chicago: Chicago University Press.
MacMillan, M. 2002. *Paris 1919: Six Months that Changed the World.* New York: Random House.
———. 2014. *The War that Ended Peace: The Road to 1914.* New York: Random House.
Majerus, M. E. N. 1998. *Melanism: Evolution in Action.* Oxford: Oxford University Press.
Malthus, T. R. 1798. *An Essay on the Principle of Population.* London: Printed for J. Johnson, In St. Paul's Church-Yard.
———. [1826] 1890. *An Essay on the Principle of Population (Sixth Edition).* London: Ward, Lock, and Co.
Marrin, A. 1974. *The Last Crusade: The Church of England in the First World War.* Durham, NC: Duke University Press.
Maurice, F. D. 1869. Lecture XI: War. *Social Morality: Twenty-One Lectures Delivered in the University of Cambridge,* 199–223. London and Cambridge: Macmillan.
Mayhew, R. J. 2014. *Malthus: The Life and Legacies of an Untimely Prophet.* Cambridge, MA: Harvard University Press.
Mayr, E. 1942. *Systematics and the Origin of Species.* New York: Columbia University Press.
McGrath, A. E., editor. 1995. *The Christian Theology Reader.* Oxford: Blackwell.
McShea, D., and R. Brandon. 2010. *Biology's First Law: The Tendency for Diversity and Complexity to Increase in Evolutionary Systems.* Chicago: University of Chicago Press.
Meyer, T. H., editor. 1997. *Light for the New Millennium: Rudolf Steiner, Helmuth von Moltke, Eliza von Moltke. Letters, Documents and After-Death Communications.* Forest Row, Sussex: Rudolf Steiner Press.
Mitman, G. 1992. *The State of Nature: Ecology, Community, and American Social Thought, 1900–1950.* Chicago: University of Chicago Press.
Montgomery, L. M. 1921. *Rilla of Ingleside.* Toronto: McClelland and Stewart.
Montagu, A. 1976. *The Nature of Human Aggression.* Oxford: Oxford University Press.
Moore, A. 1889. The Christian doctrine of God. *Lux Mundi.* Editor C. Gore. London: John Murray.
Moore, G. E. 1903. *Principia Ethica.* Cambridge: Cambridge University Press.
Morris, D. 1967. *The Naked Ape: A Zoologist's Study of the Human Animal.* New York: McGraw-Hill.

Morris, I. 2014. *War! What Is It Good For? Conflict and the Progress of Civilization from Primates to Robots*. New York: Farrar, Straus and Giroux.

Mozley, J. B. 1871. *War: A Sermon Preached Before the University of Oxford*. London: Longmans, Green.

Naden, C. 1999. *Poetical Works of Constance Naden*. Kernville, CA: High Sierra Books.

Nagel, T. 2012. *Mind and Cosmos: Why the Materialist Neo-Darwinian Conception of Nature Is Almost Certainly False*. New York: Oxford University Press.

National Conference of Catholic Bishops. 1983. *The Challenge of Peace: God's Promise and Our Response: A Pastoral Letter on War and Peace*. Washington, DC: United States Catholic Conference.

Nesse, R. M., and G. C. Williams. 1994. *Why We Get Sick: The New Science of Darwinian Medicine*. New York: Times Books.

Niebuhr, R. [1932] 2015. *Moral Man and Immoral Society*. In *Major Works on Religion and Politics*. R. Niebuhr, 135–350. New York: Library of America.

———. [1940] 2015. An end to illusions. *Major Works on Religion and Politics*. R. Niebuhr, 619–23. New York: Library of America.

———. 1941. *The Nature and Destiny of Man. I. Human Nature*. New York: Charles Scribner's Sons.

———. 1943. *The Nature and Destiny of Man. II. Human Destiny*. New York: Charles Scribner's Sons.

———. 1965. *Man's Nature and His Communities*. New York: Scribner's.

———. 1991. Why the Christian Church is not pacifist. *Reinhold Niebuhr: Theologian of Public Life*. Editor L. Rasmussen, Minneapolis, MN: Fortress Press.

———. 2015a. The bombing of Germany. *Major Works on Religion and Politics*. R. Niebuhr, 654–55. New York: The Library of America.

———. 2015b. The death of a martyr. *Major Works on Religion and Politics*. R. Niebuhr, 656–59. New York: Library of America.

Nietzsche, F. 1895 [1990]. *The Anti-Christ*. London: Penguin.

Numbers, R. L. 2006. *The Creationists: From Scientific Creationism to Intelligent Design*. Standard ed. Cambridge, MA: Harvard University Press.

O'Brien, W. V. 1992. Desert Storm: A just war analysis. *St. John's Law Review* 66: 797–823.

O'Donovan, O. 2003. *The Just War Revisited*. Cambridge: Cambridge University Press.

Olby, R. C. 1963. Charles Darwin's Manuscript of Pangenesis. *The British Journal for the History of Science* 1: 251–63.

Orend, B. 2000. *War and International Justice: A Kantian Perspective*. Waterloo, Ontario: Wilfred Laurier University Press.

Owen, R. 1848. *On the Archetype and Homologies of the Vertebrate Skeleton*. London: Voorst.

Paley, W. [1802] 1819. *Natural Theology (Collected Works: IV)*. London: Rivington.

Pavlischek, K. 2008. Reinhold Niebuhr, Christian Realism, and Just War Theory: A Critique. *Christianity and Power Politics Today: Christian Realism and Contemporary Christian Dilemmas*. E. Patterson, 53–71. New York: Palgrave Macmillan.

Pennock, R., and M. Ruse, editors. 2008. *But Is It Science? The Philosophical Question in the Creation/Evolution Controversy (Second Edition)*. Buffalo, NY: Prometheus.

Philpott, W. 2009. *Bloody Victory: The Sacrifice on the Somme*. London: Little, Brown.

Pinker, S. 2011. *The Better Angels of Our Nature: Why Violence Has Declined.* New York: Viking.

———. 2018. *Enlightenment Now: The Case for Reason, Science, Humanism, and Progress*. New York: Viking.

Plantinga, A. 1991. When faith and reason clash: evolution and the Bible. *Christian Scholar's Review* 21 (1): 8–32.

———. 2011. *Where the Conflict Really Lies: Science, Religion, and Naturalism.* New York: Oxford University Press.

Potts, M., and T. Hayden. 2010. *Sex and War: How Biology Explains Warfare and Terrorism and Offers a Path to a Safer World.* Dallas, TX: BenBella Books.

Professors of Germany. [1914] 1919. To the Civilized World. *The North American Review* 210: 284–287.

Provine, W. B. 1971. *The Origins of Theoretical Population Genetics*. Chicago: University of Chicago Press.

———. 1986. *Sewall Wright and Evolutionary Biology*. Chicago: University of Chicago Press.

Ramsey, A. M. 1960. *An Era in Anglican Theology from Gore to Temple: The Development of Anglican Theology between "Lux Mundi" and the Second World War 1889–1939.* New York: Charles Scribner's.

Ramsey, P. 1961. *War and the Christian Conscience: How Shall Modern War Be Conducted Justly*. Durham, NC: Duke University Press.

Reade, W. W. [1875] 2012. *The Outcast*. Los Angeles: IndoEuropean.

Richards, R. J. 1987. *Darwin and the Emergence of Evolutionary Theories of Mind and Behavior*. Chicago: University of Chicago Press.

———. 2003. *The Romantic Conception of Life: Science and Philosophy in the Age of Goethe*. Chicago: University of Chicago Press.

———. 2008. *The Tragic Sense of Life: Ernst Haeckel and the Struggle over Evolutionary Thought*. Chicago: University of Chicago Press.

———. 2013. *Was Hitler a Darwinian? Disputed Questions in the History of Evolutionary Theory*. Chicago: University of Chicago Press.

Richards, R. J., and M. Ruse. 2016. *Debating Darwin*. Chicago: University of Chicago Press.

Roberts, J. H. 1988. *Darwinism and the Divine in America: Protestant Intellectuals and Organic Evolution, 1859–1900*. Madison: University of Wisconsin Press.

Ruse, M. 1975. Darwin's debt to philosophy: An examination of the influence of the philosophical ideas of John F. W. Herschel and William Whewell on the development of Charles Darwin's theory of evolution. *Studies in History and Philosophy of Science* 6: 159–181.

———. 1979a. *The Darwinian Revolution: Science Red in Tooth and Claw*. Chicago: University of Chicago Press.

———. 1979b. *Sociobiology: Sense or Nonsense?* Dordrecht, Holland: Reidel.

———. 1980. Charles Darwin and group selection. *Annals of Science* 37: 615–30.

———. 1982. *Darwinism Defended: A Guide to the Evolution Controversies*. Reading, MA: Benjamin/Cummings.

———, editor. 1988. *But Is It Science? The Philosophical Question in the Creation/Evolution Controversy.* Buffalo, NY: Prometheus.

———. 1996. *Monad to Man: The Concept of Progress in Evolutionary Biology*. Cambridge, MA: Harvard University Press.
———. 2001. *Can a Darwinian be a Christian? The Struggle between Science and Religion*. Cambridge: Cambridge University Press.
———. 2003. *Darwin and Design: Does Evolution Have a Purpose?* Cambridge, MA: Harvard University Press.
———. 2004. Adaptive landscapes and dynamic equilibrium: The Spencerian contribution to twentieth-century American evolutionary biology. *Darwinian Heresies*. Editors A. Lustig, R. J. Richards, and M. Ruse, 131–150. Cambridge: Cambridge University Press.
———. 2005. *The Evolution-Creation Struggle*. Cambridge, MA: Harvard University Press.
———. 2006. *Darwinism and Its Discontents*. Cambridge: Cambridge University Press.
———, editor. 2009. *Philosophy after Darwin: Classic and Contemporary Readings*. Princeton: Princeton University Press.
———. 2010. *Science and Spirituality: Making Room for Faith in the Age of Science*. Cambridge: Cambridge University Press.
———. 2012. *The Philosophy of Human Evolution*. Cambridge: Cambridge University Press.
———. 2013. The primate who knew too much. *Planet of the Apes and Philosophy: Great Apes Think Alike*. Editor J. Huss, 153–163. Chicago: Open Court.
———. 2017a. *Darwinism as Religion: What Literature Tells Us About Evolution*. Oxford: Oxford University Press.
———. 2017b. *On Purpose*. Princeton, NJ: Princeton University Press.
Russell, B. 1916. *Principles of Social Reconstruction*. London: Allen and Unwin.
Russell, E. S. 1916. *Form and Function: A Contribution to the History of Animal Morphology*. London: John Murray.
Sassoon, S. 1917. *The Old Huntsman and Other Poems*. London: Heinemann.
Segerstrale, U. 1986. Colleagues in conflict: An in vitro analysis of the sociobiology debate. *Biology and Philosophy* 1: 53–88.
Sepkoski, D. 2012. *Rereading the Fossil Record: The Growth of Paleobiology as an Evolutionary Discipline*. Chicago: Chicago University Press.
Sepkoski, D., and M. Ruse, editors. 2009. *The Paleobiological Revolution*. Chicago: University of Chicago Press.
Shermer, M. 2002. *In Darwin's Shadow: The Life and Science of Alfred Russel Wallace*. New York: Oxford University Press.
Sheppard, H. R. L. 1935. If another war comes, what will you do? *Some of My Religion*. H. R. L. Sheppard, 54–57. London: Cassell.
Sidgwick, H. 1876. The theory of evolution in its application to practice. *Mind* 1: 52–67.
Simpson, G. G. 1944. *Tempo and Mode in Evolution*. New York: Columbia University Press.
Smocovitis, V. B. 1999. The 1959 Darwin centennial celebration in America. *Osiris* 14: 274–323.
Sober, E., and D. S. Wilson. 1998. *Unto Others: The Evolution and Psychology of Unselfish Behavior*. Cambridge, MA: Harvard University Press.
Solano, E. J. 1918. *The Pacifist Lie*. London: John Murray.

Spencer, H. 1851. *Social Statics; Or the Conditions Essential to Human Happiness Specified and the First of Them Developed*. London: J. Chapman.
———. 1852. A theory of population, deduced from the general law of animal fertility. *Westminster Review* 1: 468–501.
———. 1857. Progress: Its law and cause. *Westminster Review* LXVII: 244–267.
———. 1860. The social organism. *Westminster Review*. LXXIII: 90–121.
———. 1879. *The Data of Ethics*. London: Williams and Norgate.
———. 1882. *Political Institutions: Being Part V of the Principles of Sociology*. London: Williams and Norgate.
———. 1904. *Autobiography*. London: Williams and Norgate.
Stebbins, G. L. 1950. *Variation and Evolution in Plants*. New York: Columbia University Press.
Stevenson, D. 2004. *1914–1918: The History of the First World War*. London: Allen Lane.
———. 2011. *With Our Backs to the Wall: Victory and Defeat in 1918*. Cambridge, MA: Belknap.
Storr, A. 1968. *Human Aggression*. New York: Atheneum.
Strachan, H. 2004. *The First World War*. New York: Viking.
Sussman, R. W. 2013. Why the legend of the killer ape never dies. *War, Peace, and Human Nature: The Convergence of Evolutionary and Cultural Views*. Editor D. P. Fry, 97–111. Oxford: Oxford University Press.
Swedenborg, E. 1771 [1907]. *The True Christian Religion*. New York: Houghton Mifflin.
Temple, W. 1914. *Christianity and War*. London: Oxford University Press.
Tennyson, A. [1850] 1973. In Memoriam. *In Memoriam: An Authoritative Text Backgrounds and Sources Criticism.* Editor R. H. Ross, 3–90. New York: W. W. Norton.
Thompson, J. A. 1915. Eugenics and the War. *Eugenics Review* 7: 1–14.
Thorpe, W. H. 1961. *Biology, Psychology and Belief.* Cambridge: Cambridge University Press.
———. 1962. *Biology and the Nature of Man*. Oxford: Oxford University Press.
———. 1966. *Science, Man and Morals*. Ithaca, NY: Cornell University Press.
———. 1968. *Quakers and Humanists: Swarthmore Lecture, 1968*. London: Friends Home Service Committee.
———. 1974. *Animal Nature and Human Nature*. Garden City, NY: Anchor-Doubleday.
Tiger, L., and R. Fox. 1971. *The Imperial Animal*. New York: Dell.
Tinbergen, N. 1968. On war and peace in animals and man. *Science* 160: 1411–1418.
Titius, A. 1915. *Unser Krieg: Ethische Betrachtungen*. Tübingen: Mohr.
Todes, D. P. 1989. *Darwin without Malthus: The Struggle for Existence in Russian Evolutionary Thought*. New York: Oxford University Press.
Tolkien, J. R. R. 2005. *The Lord of the Rings—Fiftieth Anniversary Edition*. New York: Mariner Books.
Towle, P. 2010. *Going to War: British Debates from Wilberforce to Blair*. London: Palgrave Macmillan.
Toynbee, P. 1959. *The Fearful Choice*. Detroit, MI: Wayne State University Press.
Tutt, J. W. 1890. Melanism and melanochroism in British lepidoptera. *The Entomologist's Record, and Journal of Variation* 1 (3): 49–56.
Tuttle, R. H. 2014. *Apes and Human Evolution*. Cambridge, MA: Harvard University Press.

Vox, L. 2017. *Existential Threats: American Apocalyptic Beliefs in the Technological Era*. Philadelphia: University of Pennsylvania Press.
von Bernhardi, F. 1912. *Germany and the Next War*. London: Edward Arnold.
———. 1914. *Britain as Germany's Vassal*. London: Dawson.
———. 1920. *Vom Kriege der Zukunft. Nach den Erfahrungen des Weltkrieges*. Berlin: Mittler.
Wallace, A. R. 1858. On the tendency of varieties to depart indefinitely from the original type. *Journal of the Proceedings of the Linnean Society, Zoology* 3: 53–62.
———. 1864. The origin of human races and the antiquity of man deduced from the theory of natural selection. *Journal of the Anthropological Society of London* 2: clvii–clxxxvii.
———. 1899. The causes of war, and the remedies. *Clarion* July 18: 213.
———. 1905. *My Life: A Record of Events and Opinions*. London: Chapman and Hall.
Walzer, M. 1977. *Just and Unjust Wars*. New York: Basic Books.
Warr, C. L. 1960. *The Glimmering Landscape*. London: Hodder and Stoughton.
Waters, C. K., and A. van Helden, editors. 1992. *Julian Huxley: Biologist and Statesman of Science*. Houston, TX: Rice University Press.
Weidman, N. 2011. Popularizing the ancestry of man: Robert Ardrey and the killer instinct. *Isis* 102: 269–299.
Weikart, R. 2004. *From Darwin to Hitler: Evolutionary Ethics, Eugenics, and Racism in Germany*. New York: Palgrave Macmillan.
Whetham, W. C. D. 1917. *The War and the Nation*. London: John Murray.
Whewell, W. 1840. *The Philosophy of the Inductive Sciences*. London: Parker.
Whitcomb, J. C., and H. M. Morris. 1961. *The Genesis Flood: The Biblical Record and Its Scientific Implications*. Philadelphia: Presbyterian and Reformed Publishing Company.
Wilkinson, A. 1978. *The Church of England and the First World War*. London: SCM Press.
Williams, G. C. 1966. *Adaptation and Natural Selection*. Princeton, NJ: Princeton University Press.
Williams, N. P. 1927. *The Ideas of the Fall and of Original Sin*. London: Longmans, Green.
Wilson, D. S., and E. O. Wilson. 2007. Rethinking the theoretical foundation of sociobiology. *Quarterly Review of Biology* 82: 327–48.
Wilson, E. O. 1975. *Sociobiology: The New Synthesis*. Cambridge, MA: Harvard University Press.
———. 1978. *On Human Nature*. Cambridge, MA: Harvard University Press.
———. 1984. *Biophilia*. Cambridge, MA: Harvard University Press.
———. 1992. *The Diversity of Life*. Cambridge, MA: Harvard University Press.
———. 1994. *Naturalist*. Washington, DC: Island Books/Shearwater Books.
———. 1998. *Consilience*. New York: Knopf.
———. 2006. *The Creation: A Meeting of Science and Religion*. New York: W. W. Norton.
Winnington Ingram, A. F. 1914. *Drinking the Cup: A Sermon Preached by the Right Hon. and Right Rev. Arthur F. Winnington Ingram D.D., Lord Bishop of London. St. Paul's Cathedral on August 9, 1914, After the Outbreak of War*. London: Wells, Gardner, and Darton.
———. 1917. *The Potter and the Clay*. London: Wells, Gardner, and Darton.

Wrangham, R., and D. Peterson. 1996. *Demonic Males: Apes and the Origins of Human Violence*. New York: Houghton Mifflin.
Wright, Q. 1942. *A Study of War*. Chicago: University of Chicago Press.
Wright, S. 1931. Evolution in Mendelian populations. *Genetics* 16: 97–159.
———. 1932. The roles of mutation, inbreeding, crossbreeding and selection in evolution. *Proceedings of the Sixth International Congress of Genetics* 1: 356–366.
Zaleski, P., and C. Zaleski. 2015. *The Fellowship: The Literary Lives of the Inklings: J.R.R. Tolkien, C. S. Lewis, Owen Barfield, Charles Williams*. New York: Farrar, Straus and Giroux.

INDEX